Xing Cai and T.-C. Jim Yeh (Eds.)

Quantitative Information Fusion for Hydrological Sciences

Studies in Computational Intelligence, Volume 79

Editor-in-chief
Prof. Janusz Kacprzyk
Systems Research Institute
Polish Academy of Sciences
ul. Newelska 6
01-447 Warsaw
Poland
E-mail: kacprzyk@ibspan.waw.pl

Further volumes of this series
can be found on our homepage:
springer.com

Vol. 58. Jeng-Shyang Pan, Hsiang-Cheh Huang, Lakhmi C. Jain and Wai-Chi Fang (Eds.)
Intelligent Multimedia Data Hiding, 2007
ISBN 978-3-540-71168-1

Vol. 59. Andrzej P. Wierzbicki and Yoshiteru Nakamori (Eds.)
Creative Environments, 2007
ISBN 978-3-540-71466-8

Vol. 60. Vladimir G. Ivancevic and Tijana T. Ivacevic
Computational Mind: A Complex Dynamics Perspective, 2007
ISBN 978-3-540-71465-1

Vol. 61. Jacques Teller, John R. Lee and Catherine Roussey (Eds.)
Ontologies for Urban Development, 2007
ISBN 978-3-540-71975-5

Vol. 62. Lakhmi C. Jain, Raymond A. Tedman and Debra K. Tedman (Eds.)
Evolution of Teaching and Learning Paradigms in Intelligent Environment, 2007
ISBN 978-3-540-71973-1

Vol. 63. Wlodzislaw Duch and Jacek Mańdziuk (Eds.)
Challenges for Computational Intelligence, 2007
ISBN 978-3-540-71983-0

Vol. 64. Lorenzo Magnani and Ping Li (Eds.)
Model-Based Reasoning in Science, Technology, and Medicine, 2007
ISBN 978-3-540-71985-4

Vol. 65. S. Vaidya, L.C. Jain and H. Yoshida (Eds.)
Advanced Computational Intelligence Paradigms in Healthcare-2, 2007
ISBN 978-3-540-72374-5

Vol. 66. Lakhmi C. Jain, Vasile Palade and Dipti Srinivasan (Eds.)
Advances in Evolutionary Computing for System Design, 2007
ISBN 978-3-540-72376-9

Vol. 67. Vassilis G. Kaburlasos and Gerhard X. Ritter (Eds.)
Computational Intelligence Based on Lattice Theory, 2007
ISBN 978-3-540-72686-9

Vol. 68. Cipriano Galindo, Juan-Antonio Fernández-Madrigal and Javier Gonzalez
A Multi-Hierarchical Symbolic Model of the Environment for Improving Mobile Robot Operation, 2007
ISBN 978-3-540-72688-3

Vol. 69. Falko Dressler and Iacopo Carreras (Eds.)
Advances in Biologically Inspired Information Systems: Models, Methods, and Tools, 2007
ISBN 978-3-540-72692-0

Vol. 70. Javaan Singh Chahl, Lakhmi C. Jain, Akiko Mizutani and Mika Sato-Ilic (Eds.)
Innovations in Intelligent Machines-1, 2007
ISBN 978-3-540-72695-1

Vol. 71. Norio Baba, Lakhmi C. Jain and Hisashi Handa (Eds.)
Advanced Intelligent Paradigms in Computer Games, 2007
ISBN 978-3-540-72704-0

Vol. 72. Raymond S.T. Lee and Vincenzo Loia (Eds.)
Computation Intelligence for Agent-based Systems, 2007
ISBN 978-3-540-73175-7

Vol. 73. Petra Perner (Ed.)
Case-Based Reasoning on Images and Signals, 2008
ISBN 978-3-540-73178-8

Vol. 74. Robert Schaefer
Foundation of Global Genetic Optimization, 2007
ISBN 978-3-540-73191-7

Vol. 75. Crina Grosan, Ajith Abraham and Hisao Ishibuchi (Eds.)
Hybrid Evolutionary Algorithms, 2007
ISBN 978-3-540-73296-9

Vol. 76. Subhas Chandra Mukhopadhyay and Gourab Sen Gupta (Eds.)
Autonomous Robots and Agents, 2007
ISBN 978-3-540-73423-9

Vol. 77. Barbara Hammer and Pascal Hitzler (Eds.)
Perspectives of Neural-Symbolic Integration, 2007
ISBN 978-3-540-73953-1

Vol. 78. Costin Badica (Ed.)
Intelligent and Distributed Computing, 2008
ISBN 978-3-540-74929-5

Vol. 79. Xing Cai and T.-C. Jim Yeh (Eds.)
Quantitative Information Fusion for Hydrological Sciences, 2008
ISBN 978-3-540-75383-4

Xing Cai
T.-C. Jim Yeh
(Eds.)

Quantitative Information Fusion for Hydrological Sciences

With 81 Figures and 7 Tables

Springer

Xing Cai
Simula Research Laboratory
P.O. Box 134
1325 Lysaker
Norway
xingca@simula.no

and

Department of Informatics
University of Oslo
P.O. Box 1080 Blindern
0316 Oslo
Norway

T.-C. Jim Yeh
Department of Hydrology
and Water Resources
The University of Arizona
Tucson, Arizona 85721
USA
ybiem@mac.hwr.arizona.edu

ISBN 978-3-642-09461-3 e-ISBN 978-3-540-75384-1

Studies in Computational Intelligence ISSN 1860-949X

Softcover reprint of the hardcover 1st edition 2008

Cover design: WMX Design GmbH, Heidelberg

Printed on acid-free paper

9 8 7 6 5 4 3 2 1

springer.com

Preface

Water is vital for the Earth and all the life forms on it, thus the importance of hydrology–the science of water–goes without saying. Water resources involve interplay between geologic, hydrologic, chemical, atmospheric, and biological processes. To study the occurrence, movement, distribution, and quality of water throughout our globe is clearly a challenging task, which requires a joining force between the different branches of hydrology, from hydrometeorology, surface water hydrology to hydrogeology and hydrochemistry as well as hydrogeophysics and hydroecology, to just mention a few. Besides a conceptual understanding, quantitative monitoring, characterization, predictions and management must resort to collaborative mathematical models and numerical algorithms, often together with computer simulations.

In recent years, massive amounts of high-quality hydrologic field data are being collected at various spatial-temporal scales using a variety of new techniques. Availability of these massive amounts of data has begun to call for a quantitative integration of geologic, hydrologic, chemical, atmospheric, and biological information to characterize and predict natural systems in hydrological sciences. Intelligent computation and information fusion as such become a key to the future hydrological sciences. We envision this subject to become a new research field that will dramatically improve the traditional approach of only qualitatively characterizing natural systems.

This edited volume contains eight chapters written by some of the leading researchers in hydrological sciences. The chapters address some of the most important ingredients for quantitative hydrological information fusion. The book aims to provide both established scientists and graduate students with a summary of recent developments in this new research direction, while shedding some light into the future.

The eight chapters can be divided into three mutually overlapping parts. The first part consists of Chapters 1 and 2 which mainly address the methodological issues. In particular, Chapter 1 discusses different data fusion techniques for integrating hydrological models, where the discussion is carried out from the perspective of hydroinformatics and computational intelligence.

Chapter 2 depicts an advanced computational environment that enables interactive and real-time 3D groundwater modeling. The combined power of parallel computing, dynamic visualization, and computational steering enables a fusion of flow modeling, transport modeling, subscale modeling, uncertainty modeling, geostatistical simulation, and GIS mapping.

As the second theme of the book, Chapters 3-6 concentrate on some mathematical and numerical methods. Using the Kalman filter based on Karhunen-Loève decomposition, the authors of Chapter 3 show how to reduce the uncertainty in characterizing hydraulic medium properties and system responses. Chapter 4 presents efficient data analysis tools using trajectory-based methods, which also offer insight into inverse modeling of flow and transport. In close relation, Chapter 5 describes streamline methods that are capable of reconciling 3D geological models to dynamic reservoir responses. Another numerical technique in inverse modeling is given in Chapter 6, which addresses a systematic regularized inversion approach to incorporating geophysical information into the analysis of tomographic pumping tests.

The third part of the present book focuses on real-life applications of hydrological information fusion. Chapter 7 is about using satellite rainfall datasets and hydrologic process controls for flood prediction in ungauged basins, whereas Chapter 8 reports an engineering case of groundwater management by integrating large-scale zoning of aquifer parameters and a sedimentary structure-based heterogeneous description of the aquifer properties.

The idea of the present book was conceived following a warm suggestion by Prof. Dr. Janusz Kacprzyk, Series Editor of Studies in Computational Intelligence at Springer. We are therefore greatly indebted to Prof. Kacprzyk for his advice and encouragement. Engineering Editor Dr. Thomas Ditzinger and Heather King at Springer's Engineering Editorial Department, in particular, deserve our sincere thanks for their patient guidance and technical support throughout the editorial process. We are of course tremendously grateful to all the contributed authors for carefully preparing their chapters. Moreover, positive response from numerous researchers to our call-for-chapters is acknowledged, although they were not able to contribute in the end due to the tight time schedule.

Last but not least, we wish to express our heartfelt gratitude to a large number of anonymous reviewers, who carefully read through the earlier versions of the book chapters and provided valuable suggestions for improvement. There is no exaggeration in saying that this book project has been a team work from start to finish. We sincerely hope that this book will give the reader an equal amount of pleasure as it has given us during the editing work.

Oslo & Tucson, July 2007

Xing Cai
T.-C. Jim Yeh

List of Contributors

Geoffrey C. Bohling
Kansas Geological Survey
University of Kansas
Lawrence, KS 66047
USA
geoff@kgs.ku.edu

Yan Chen
Mewbourne School of Petroleum
and Geological Engineering
University of Oklahoma
Norman, OK 73019
USA
yan@ou.edu

Hao Cheng
Department of Petroleum
Engineering
Texas A&M University
College Station, TX 77843
USA
hao.cheng@chevron.com

Akhil Datta-Gupta
Department of Petroleum
Engineering, Texas A&M
University College Station
TX 77843
USA
datta-gupta@tamu.edu

Deepak Devegowda
Department of Petroleum
Engineering, Texas A&M University
College Station, TX 77843
USA
deepak.devegowda@pe.tamu.edu

Jannis Epting
Department of Environmental
Sciences, Applied and
Environmental Geology
University of Basel, 4056 Basel
Switzerland
jannis.epting@unibas.ch

Faisal Hossain
Department of Civil
and Environmental Engineering
Tennessee Technological University
Cookeville, TN 38505
USA
fhossain@tntech.edu

Peter Huggenberger
Department of Environmental
Sciences, Applied
and Environmental Geology
University of Basel, 4056 Basel
Switzerland
peter.huggenberger@unibas.ch

Nitin Katiyar
Department of Civil
and Environmental Engineering
Tennessee Technological University
Cookeville, TN 38505
USA
nitin.katiyar@gmail.com

Ralph Kirchhofer
Department of Environmental
Sciences, Applied
and Environmental Geology
University of Basel, 4056 Basel
Switzerland
ralph.kirchhofer@unibas.ch

Shu-Guang Li
Department of Civil
and Environmental Engineering
Michigan State University
East Lansing, MI 48824
USA
lishug@egr.msu.edu

Qun Liu
Department of Civil
and Environmental Engineering
Michigan State University
East Lansing, MI 48824
USA
liuqu@egr.msu.edu

Zhiming Lu
Earth and Environmental
Sciences Division
Los Alamos National Laboratory
Los Alamos, NM 87645
USA
zhiming@lanl.gov

Dayo Oyerinde
Department of Petroleum
Engineering
Texas A&M University
College Station, TX 77843
USA
dayo.oyerinde@pe.tamu.edu

Christian Regli
GEOTEST AG, 7260 Davos Dorf
Switzerland
christian.regli@geotest.ch

Linda See
School of Geography
University of Leeds Leeds
LS2 9JT
UK
l.m.see@leeds.ac.uk

Natalie Spoljaric
Department of Environmental
Sciences, Applied and Environmental
Geology University of Basel
4056 Basel
Switzerland
natalie.spoljaric@stud.
unibas.ch

Donald W. Vasco
Lawrence Berkeley National
Laboratory
University of California
Berkeley, CA 94720
USA
dwvasco@lbl.gov

Dongxiao Zhang
Department of Civil and Environmental Engineering and Mork Family
Department of Chemical Engineering
and Material Sciences
University of Southern California
Los Angeles, CA 90089
USA
donzhang@usc.edu

Contents

Data Fusion Methods for Integrating Data-driven Hydrological Models

Linda See

School of Geography, University of Leeds

Summary. This chapter will address the use of different data fusion techniques for integrating or combining hydrological models. Different approaches will be demonstrated using flow forecasting models from the River Ouse catchment in the UK for a lead time of 6 hours. These approaches include simple averaging, neural networks, fuzzy logic, M5 model trees and instance-based learning. The results show that the data fusion approaches produce better performing models compared to the individual models on their own. The potential of this approach is demonstrated yet remains largely unexplored in real-time hydrological forecasting.

1 Introduction

Approaches to hydrological models are varied and lie on a spectrum that characterises the degree to which they encapsulate physical processes. On one end of the scale are fully distributed physical models based on the laws of the conservation of energy and mass (e.g. the SHE model of Abbott et al. (1986)). Conceptual models fall in the middle of the spectrum as parameterisation increases while the opposite end is dominated by data driven models (DDM) or what Wheater et al. (1993) refer to as metric models. As the name suggests, DDM is based on finding relationship between the input and output variables of a system without explicit knowledge of its physical behaviour. Physical models have their limitations because many of the hydrological processes are complex and difficult to represent. Understanding of the system is also far from complete so DDM offers an alternative approach to traditional physically-based models. DDM has been the subject of much research activity in hydrological modelling over the two last decades and includes a range of different techniques. These mainly originate from the fields of computational and artificial intelligence (Solomatine, 2005), and include techniques such as neural networks (NN), fuzzy logic, evolutionary computing and machine learning. Many

L. See: *Data Fusion Methods for Integrating Data-driven Hydrological Models*, Studies in Computational Intelligence (SCI) **79**, 1–18 (2008)
www.springerlink.com

of these approaches have shown promising results for hydrological modelling; some examples include:

- flood prediction (Minns and Hall, 1996; Smith and Eli, 1995; Shamseldin, 1997; Khondker et al., 1998; Salas et al., 2000; Babovic and Keijzer, 2002; Jain and Srinivasulu, 2004; Nayak et al., 2004; Heppenstall et al., 2006)
- urban stormwater runoff (di Pierro et al., 2006)
- irrigation (Yang et al., 1997; Schmitz et al., 2002)
- water quality prediction (Maier and Dandy, 1996; Gumrah et al., 2000; Brion et al., 2001; Ha et al., 2003; Chaves and Chang, 2006)
- groundwater prediction and protection (Gangopadhyay *et al.*, 1999; Lindsay et al., 2002; Singh et al., 2003; Giustolisi and Simeone, 2006; Yan and Minsker, 2006)
- reservoir operation (Kojiri and Hanatani, 2002; Chaves et al., 2003)
- the potential effects of climate change on the water system (Clair and Ehrman, 1998; Kojiri, 1999).

Attempts by Young (2002, 2003) to link conceptual and data-driven models in their data-based mechanistic approach are also proving very useful for flood forecasting. Wheater et al. (1993) refer to this approach as hybrid metric-conceptual models, which may appeal to hydrologists interested in both performance and improvement of physical understanding.

Many of the data-driven hydrological models found in the literature are examples of global models. However, the physical processes underlying these models are complex and the development of a single global solution may not be the best approach. An alternative is the development of several individual models, which are then combined using a data fusion strategy. Data fusion can be defined as the process of combining or integrating information from multiple sensors and/or data sources in order to provide a solution that is either more accurate according to some measure of performance or which allows one to make additional inferences above and beyond those which could be achieved through the use of single source data alone (Dasarathy, 1997). Just as one is able to draw links between methods of artificial intelligence and the biological system or principle upon which they are loosely based, the general concept of data fusion is analogous to the manner in which humans and animals use a combination of multiple senses, experience and the ability to reason to improve their chances of survival.

Different data fusion strategies for integrating individual rainfall-runoff models for the River Ouse catchment are presented, which include simple averaging, and the use of NNs, fuzzy logic, M5 model trees and instance-based learning. The models are assessed using global goodness-of-fit statistics and examination of the forecast hydrographs. The chapter concludes with a discussion of the potential for real-time hydrological forecasting.

2 Review of Data Fusion and Hydrological Modelling

Data fusion is a growing area of research that covers military applications such as target recognition and tracking as well as non-military ones such as law enforcement, medical diagnosis and electromechanical system diagnosis (Hall, 1992; Hall and Linas, 2001). Examples of data fusion applied to the management and operation of hydrological systems are less common. Yet these systems may benefit from data fusion, especially if different forecasting models are able to capture some aspect of the hydrological record better than another. In this way the strengths of each individual approach are exploited and a potentially better overall solution is produced.

There are many different data fusion algorithms available including Bayesian inference, Dempter-Shafer theory, neural networks and rule-based reasoning systems (Hall and Linas, 1997). Abrahart and See (2002) used neural networks and fuzzy logic to combine a set of individual rainfall-runoff models for the Rivers Ouse and Wye in the UK. Shamseldin et al. (2002) combined five rainfall-runoff models for eight catchments using neural networks, trained with a range of different transfer functions. Both studies found that the combination model outperformed the individual rainfall-runoff models.

Data fusion also appears in the hydrological literature under different names, e.g. stacking, bagging, boosting, committee machines and ensemble modelling (Breiman, 1996; Anctil and Lauzon, 2004; Solomatine and Siek, 2006). However, these names generally refer to the development of a pool of models using the same technique, e.g. a series of neural networks, which are then aggregated to produce a single model output. Stacking or ensemble modelling is most closely related to the way in which data fusion is interpreted in this chapter. This is an area where recent research attention has been focused. One of the motivations for this research is to reduce the uncertainty of the predictions that is partly a function of model structure. This is discussed by Keijzer and Babovic (2000) where the authors have decomposed the error term into two main components: error caused by bias, i.e. the ability of a model to fit a given data set, and error caused by variance, i.e. the variability associated with a particular method. When several models are fused into a single predictor, the variance error is effectively eliminated with no change in the error due to bias and a result in overall improvement in performance. For example, Georgakakos et al. (2004) examined the performance of ensembles of single distributed and lumped models from the Distributed Model Intercomparison Project (DMIP) using simple averaging compared to individual model performance. The results showed that the multimodel ensemble outperformed all individual models and should be considered seriously as tools for operational forecasting. Ajami et al. (2006) extended this work by considering different types of model integration schemes such as weighted averaging and superensembles. They found similar results, i.e. the superiority of the multimodel approach when compared to individual models, particular when introducing a bias correction to the aggregation procedure. Similar results are being reported in other multimodel ensemble studies of different catchments (Xiong et al., 2001; Marshall et al., 2006; Oudin et al., 2006). Bagging employs bootstrapping to create several new training data sets from the original data (Breiman, 1996). Neural networks (or other model types) are then trained on each of the new data sets and combined using some form of aggregation such as an average. Boosting also involves the creation of several individual models in a two step procedure. The individual models are first trained assuming that all inputs have equal effect in terms of weight updating of the model (in the case of a neural network). In the second step the input data that have resulted in a poorer result will then have a greater impact on the weight optimization. Anctil and Lauzon (2004) compare the use of stacking, as well as bagging and boosting along with two other neural network generalization methods to develop rainfall-runoff models for six catchments. They found that all the multi-model methods resulted in improved performance when compared to individual global models.

Committee machines provide another term for the combination of outputs from multiple models (Haykin, 1999). Solomatine and Siek (2006) offer a method for classifying committee machines based on how the original input data are partitioned. If the data are not split into subsets but different models are trained on the entire data set, then the approach is that of stacking, ensembles or data fusion as used in this chapter. If the data are split before training, then the committee machine contains schemes such as bagging, boosting and classification. Although many different terms exist that capture the idea of combining multiple models into a single better performing predictor, it is possible to slot the approaches used here into a larger framework that defines a hierarchy of data fusion architectures, which is presented in the next section.

3 A Framework for Data Fusion

Data fusion has developed as a result of advances in technology, i.e. new sensors that can provide varied and complex real-time information, and a powerful driver, i.e. military applications such as target tracking and recognition. The literature reveals that there is a lack of standardisation in the terminology used to describe individual applications; this is reflected in the large number of architectures that have been proposed, as pointed out by Smith and Singh (2006). However, these authors examine the model developed by the American Joint Directors of Laboratories Data Fusion Subpanel (US Department of Defense, 1991), which provides one framework that is commonly used. There are four levels in the model, which are arranged in a hierarchy:

- level 1: object refinement;
- level 2: situation assessment;
- level 3: threat assessment; and
- level 4: process assessment.

Although these terms have strong military connections, it is level 1 or object refinement into which the idea of integrating multiple models falls. This level is concerned with taking information from disparate sensors and combining them using a number of different algorithms. Singh and Smith (2006) review a diverse set of algorithms including neural networks, nearest neighbours, fuzzy logic, Dempster-Shafer rules of combination, expert systems, etc., depending upon the type of object refinement exercise. This level 1 architecture is similar to the data-in-data-out (DIDO) architecture suggested by Dasarathy (1997), which is the simplest data fusion strategy in his hierarchical classification.

Moving beyond object refinement or the DIDO architectures, data fusion can operate at higher levels using any combination of input types to produce a numerical output, a feature output or a higher level decision in the form of an expert system. For example, forecasting model outputs might be combined with expert knowledge about likelihood of flooding and the types of operational activities that should be implemented to issue warnings. It is also possible to add expert knowledge to the data fusion process via self-adaptation. Feedback loops inserted between the system output and the combination algorithm can be used to adjust or recalibrate systems in real-time, allowing for system change to be incorporated. At present the simplest data fusion techniques are not being used operationally. However, higher level data fusion implementations incorporating either decisions or adaptation may one day have significant implications for the design and construction of automated real-time flood prediction and flood warning systems.

4 Study Area

This study is undertaken on a single catchment but serves to illustrate the methodology in this chapter. Applicability of these approaches to other catchments is required before it is clear how transferable these techniques are. However, the literature review has demonstrated that multimodel approaches are showing superior performance over individual models in every reported case.

The River Ouse catchment is located in northern England (Figure 1). It has three main tributaries: the Swale, the Ure and the Nidd, and one main urban area: the City of York. The average annual rainfall is 906 mm and the average annual runoff is 464 mm (1969–90). The catchment consists of a mix of geologies: Carboniferous Limestone and Millstone Grit in the Pennine Hills and Yorkshire Dales to the west; Permo-Triassic rocks in the lowland regions to the east. Elevation ranges from 710 m above sea level at the headwaters to 10 m at the catchment outlet, situated in the Vale of York. Floods are generated most frequently in the upland area and some of the largest floods have been the result of a combination of high rainfall and snowmelt. This catchment is very responsive to rainfall and has a small baseflow component (Kuchment et al., 1996). Flooding can be a serious problem for the city of York. November

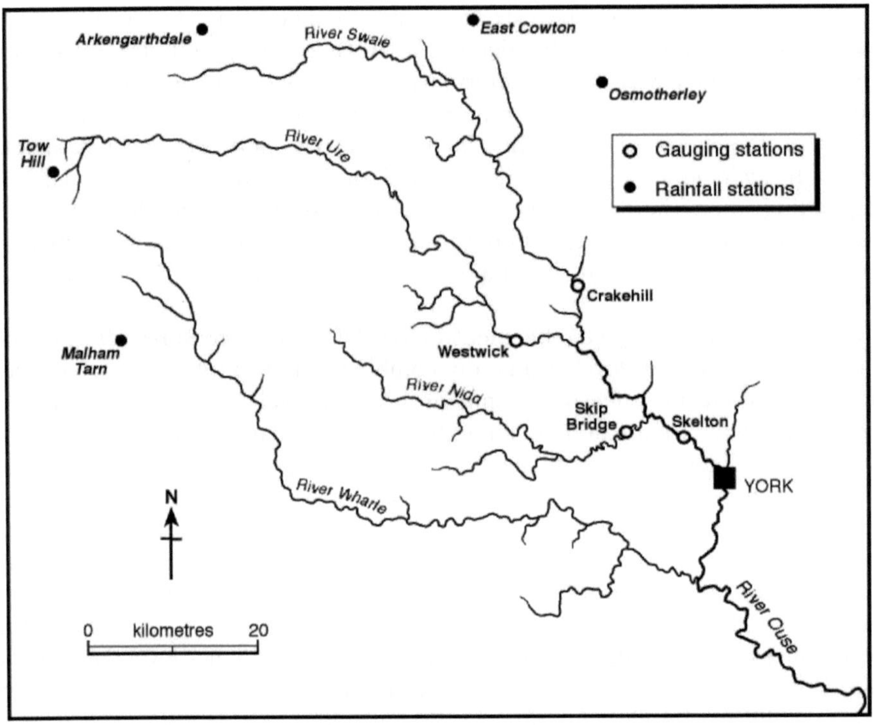

Fig. 1. Map of the River Ouse catchment.

2000 witnessed some of the worst floods on record and water levels in York were amongst the highest ever measured. Historical analysis illustrates that there was a dramatic rise in the number of floods during the 1940s, with a continual increase after that, except during the 1970s (Lane, 2002). Not only are floods becoming more frequent, but there is also a general increase in their magnitude, which may be due to land use changes (Longfield, 1998). Measurements at Skelton, a hydrological gauging station situated 5.5 km upstream of York, forms one of the stations at which operational decisions are made with respect to York. The station has a downstream location, far from the headwaters, with an upstream catchment area of 3315 km^2. More detailed information about the catchment can be found in Jarvie et al. (1997) and Law et al. (1997).

5 Application of Data Fusion Methods

Models for predicting river levels at Skelton have been developed using data from Skelton, three upstream gauging stations (Crakehill, Skip Bridge and Westwick) and average rainfall at upstream rain gauges (Tow Hill, Arkengartdale, East Cowton, Osmotherly and Malham Tarn) for a lead time of

6 hours. The input data set was split into a training data set (Jan 1989 to Jun 1991) using 60% of the data and a second independent test data set (Jul 1991 to Dec 1992) using the remaining 40%.

5.1 Development of Individual Models

Four individual models have been developed, which were then used as inputs to the data fusion models:

- A hybrid neural network
- A fuzzy logic rule-based model
- An ARMA time series model
- A model of persistence.

These individual models have been developed previously in See and Opensaw (1999; 2000) and will be described briefly in the sections that follow.

Hybrid Neural Network

Neural networks are a type of biologically inspired computational model, based loosely on the functioning of the human brain. There are many books and papers on this topic; see Haykin (1999) for further information.

The hybrid neural network (HNN) developed in See and Openshaw (1999) consists of five individual feedforward neural networks that were trained on five subsets of the data. The inputs to these networks were: the level at the current time and the previous 11 hours, 6 hours of levels at three upstream stations, lagged to reflect travel time, and 7 days of average rainfall from nearby rain gauges, resulting in a total of 37 inputs per network. The number of weights that were trained per network was 228. These subsets were created by partitioning the data into hydrological events using a self-organising map (Kohonen, 1984), producing distinct behaviours such as low levels, rising levels, falling levels, etc. Once the neural networks were trained, they were combined into a single hybrid model via a simple fuzzy logic controller. The fuzzy logic model determined which weighted combination of network predictions to use for a given set of river level conditions, and was developed using a genetic algorithm to optimise the membership functions and rulebase.

A Fuzzy Logic Rule-Based Model

As with NNs, fuzzy logic models also perform an input-output mapping but use a fuzzy inference procedure. The input and output variables are partitioned into overlapping membership functions and the model is defined by an IF-THEN fuzzy rulebase. Execution of the model produces a fuzzy solution surface that is converted back into a single numerical output value using a defuzzification algorithm (Jang et al., 1997).

The model inputs to the fuzzy logic model developed in See and Openshaw (1999) were (i) change in river level over the past 6 hours; (ii) the current daily rainfall; and (iii) the river level at three upstream stations (appropriately lagged for average travel times). In order to minimize the problem of the curse of dimensionality (Kosko, 1992), the number of variables and fuzzy set partitions was kept to a minimum. The output variable was the change in river level in six hours time, and was partitioned into nine fuzzy sets to cover the solution space adequately. A genetic algorithm was used to train a global fuzzy model on an equal proportion of the different occurrences of the five event types described in the previous section. In total there were 32 rules to calibrate. As a result, only 30% of the data were used in model development, reducing the potential bias caused by a large number of low level events in the historical record. The rulebase and fuzzy membership functions of the model can be found in See and Openshaw (2000).

An ARMA Model

An AutoRegressive Moving Average (ARMA) model (Box and Jenkins, 1976), which uses a weighted linear combination of previous values and errors in prediction to produce a forecast, was developed to predict the differences between the current river level and the value expected in 6 hours time. The models were fit to the first 60% of the data set and tested on the remaining 40%. The best fitting model had one autoregressive and one moving average term, i.e. ARMA[1,1] so there were only two weights to determine.

A Model of Persistence

A model of persistence or a naïve model substitutes the last observed value for the current prediction. It serves as a good benchmark when comparing model performance. There may also be times when the naïve model produces the best prediction and may therefore be of assistance in a data fusion approach.

5.2 Data Fusion Methods

The different data fusion strategies used in this chapter include simple arithmetic averaging, neural network and fuzzy logic models, M5 model trees and instance-based learning.

Previously Employed Methods

In Abrahart and See (2002), six different data fusion strategies were used, which were labeled DF1 to DF6. The first approach was a simple averaging of the individual models (DF1). In the second approach, the best individual performing model in the previous time step was used to make the prediction at the next time step (DF2). DF3 and DF4 used neural networks to combine the individual models, where DF3 predicts the absolute river level while DF4 predicts the change in river level. The final two approaches (DF5 and DF6) use fuzzy logic models for data fusion. DF5 uses a fuzzy logic model in which more than one individual modeling solution could be recommended at each given moment and to varying degrees. The resulting forecast is therefore a weighted average of one or more single model recommendations. The inputs to the model are the current level and the change in level over the last 6 hours. The weightings are based on the membership functions. DF6, on the other hand, is a fuzzy model that uses the current level and the forecast error at the last time step to make its recommendations. The results from these models are reported in this chapter as a comparison against the two data-driven approaches described in the following sections.

M5 Model Trees (M5)

M5 model trees are a machine-learning technique that divides the solution space into subspaces and then builds a separate regression equation for each one (Quinlan, 1992). The approach is based on information theory where the criteria for partitioning into subspaces are based on entropy measures. On a subspace scale the relationships between inputs and outputs are linear but when combined together result in the ability to capture non-linear relationships. The result is a series of rules governing the input variables and regression equations at the leaves of the tree.

M5 model trees have some important advantages over neural networks including faster development time and results that can be easily understood by decision makers. These advantages have been exploited in a recent paper in which M5 model trees were applied to a hydrological modeling problem for the Sieve catchment in Italy (Solomatine and Dulal, 2003). M5 model trees were developed to forecast discharge at one and three hours ahead using a combination of discharge at the gauging station and effective precipitation. The M5 model tree for the one hour ahead predictions essentially divided the solution space into low, medium and high flows, fitting a linear regression equation to each subset of the solution space. The results were comparable for both lead times to a neural network.

Instance-Based Learning

Instance-based learning involves combining instances from the training data set that are close in attribute space to the input vector for which a prediction is required (Solomatine, 2005). No model is actually trained or calibrated during this process. The nearest neighbour classifier is an example of the simplest instance-based learning approach and is one of the easiest to apply. This can be generalized to the k-nearest neighbour (k-NN) method, where the user specifies the value of k. A simple average is then used to combine the instances or alternatively, a weighted distance decay function can be applied. Solomatine (2005) argues that comparable results can be achieved with instance-based learning and NNs. Some successful applications of instance-based learning in hydrology include the works by Karlsson and Yakowitz (1987), Shamseldin and O'Connor (1996) and Toth et al. (2000).

Both the M5 model trees and the instance-based learning approaches have been implemented using the WEKA software, which is available from the following website: http://www.cs.waikato.ac.nz/~ml/weka/.

5.3 Evaluation of the Models

The different models are compared using three global goodness-of-fit statistics:

- Root Mean Square Error (RMSE) in meters:

$$RMSE = \sqrt{N^{-1}\sum_{i=1}^{N}(O_i - P_i)^2}$$

- Mean Absolute Error (MAE) in meters:

$$MAE = N^{-1}\sum_{i=1}^{N}|O_i - P_i|$$

- Nash-Sutcliffe Coefficient of Efficiency (CE):

$$CE = 1 - \frac{\sum_{i=1}^{N}(O_i - P_i)^2}{\sum_{i=1}^{N}(O_i - \overline{O})^2}$$

- Persistence Index (PI) (Kitanidis and Bras, 1980):

$$PI = 1 - \frac{\sum_{i=1}^{N-L}(O_{i+L} - P_{i+L})^2}{\sum_{i=1}^{N-L}(O_{i+L} - O_i)^2}$$

where O_i is the observed value at time i, P_i is the predicted value at time i, N is the total number of observations, $\overline{O}$ is the mean of O over N and L is the lead time. For the most recently developed models, these statistics have been calculated using the Hydrotest website (Dawson et al., 2007). Hydrographs are also examined from the independent test data set.

6 Comparison of Results

Table 1 contains the performance measures for the different models. It can be seen that all the data fusion models, with the exception of a simple average, outperform the individual models in all four goodness-of-fit statistics. The best performing models are DF3 and DF4, which involved integrating the individual models using a neural network, predicting either absolute or differenced values. The next best performers of the different data fusion methods are the instance-based learning and the M5 model trees. Interestingly, the model using 10 nearest neighbours did not perform as well as the one with only 5. In terms of the RMSE and MAE, which provide absolute measures of error, the individual models ranged between 55.5 to 15.9 cm for RMSE and 2.7 and 6.6 cm for MAE, where flood heights of beyond 6 metres have been recorded at this site and an average across all observations is approximately 1 metre. Compared to the averaging approach (DF1), the individual HNN model performs better. However, the RMSE of all the other data fusion approaches ranges between 1.5 to 4.9 cm for RMSE and 0.6 to 2.2 cm for MAE, indicating a substantial improvement considering that these measures take all the data points into account. The CE values are already high for the individual models but further improvements can be seen when the data fusion approaches are applied. A more interesting measure is the PI, which indicates how well a given model performs relative to the naïve or persistence model. The closer the value is to 1, the better the given model performs. Both the individual fuzzy model and the averaging of models (DF1) produced the lowest PI values. Both DF3 and DF4 have the highest PIs, which correlates with their superior results in RMSE, MAE and CE. However, all the other

Table 1. Goodness-of-fit evaluation measures for the individual and combined models for the independent test data set.

		RMSE	MAE	CE	PI
Individual Models	HNN	0.0553	0.0268	0.9947	0.879
	ARMA	0.0794	0.0338	0.9890	0.751
	Fuzzy Model	0.1091	0.0427	0.9793	0.529
	Persistence	0.1590	0.0657	0.9560	0.000
Data Fusion Models	DF1	0.0860	0.0340	0.9871	0.707
	DF2	0.0416	0.0161	0.9970	0.932
	DF3	0.0160	0.0067	0.9996	0.990
	DF4	0.0146	0.0064	0.9996	0.992
	DF5	0.0489	0.0219	0.9958	0.905
	DF6	0.0399	0.0171	0.9972	0.937
	M5	0.0394	0.0157	0.9974	0.939
	IBL 5NN	0.0376	0.0160	0.9976	0.944
	IBL 10NN	0.0387	0.0163	0.9975	0.941

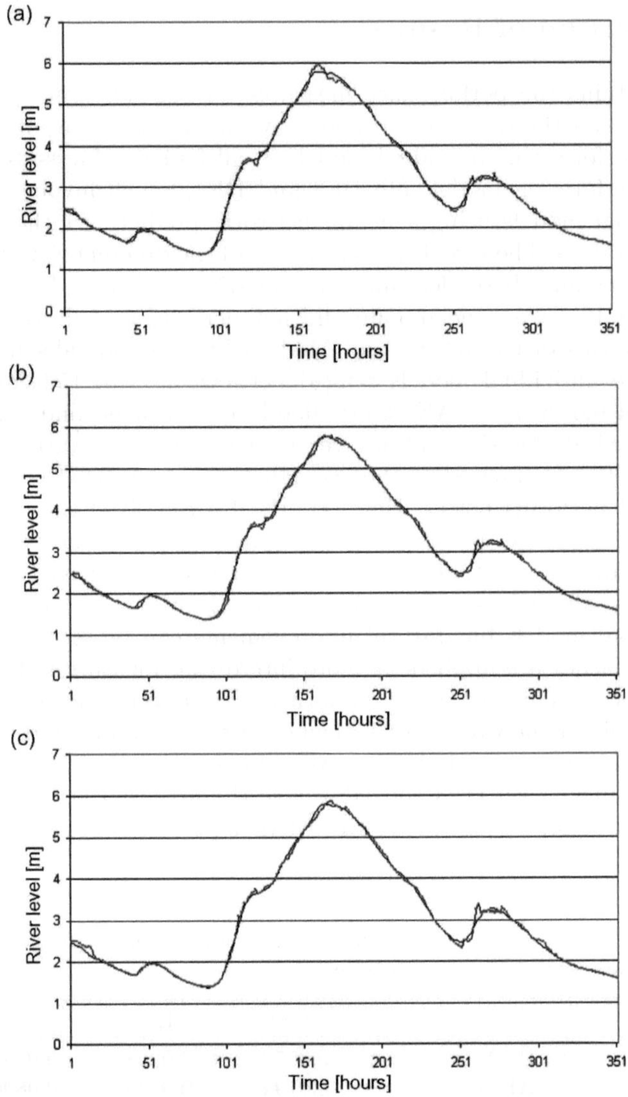

Fig. 2. Hydrograph of a storm event in the independent data set (26/11/92 to 11/12/92). The solid black line is the actual river level data while the grey line is the prediction from the (a) IBL-5 (b) IBL-10 and (c) M5 model tree.

data fusion approaches except for DF1 (averaging) have PI values close to 1, indicating that they are significantly better than the naïve or persistence model.

Figures 2 and 3 are two storm events taken from the independent data set. The storm in Figure 2 is the largest event in this data set while Figure 3 represents a more typical and frequent storm event. The figures show plots of

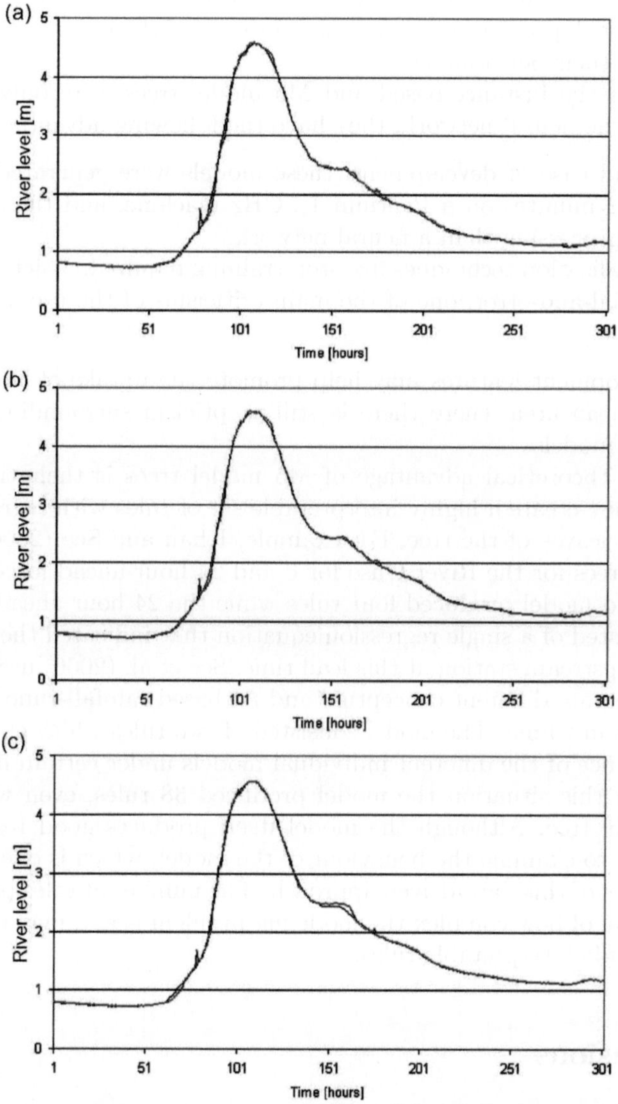

Fig. 3. Hydrograph of a storm event in the independent data set (27/03/92 to 09/04/92). The solid black line is the actual river level data while the grey line is the prediction from the (a) IBL-5 (b) IBL-10 and (c) M5 model tree.

actual level against predictions from the two instance-based learning models (a and b) and the M5 model tree (c). It is clear from the hydrographs that the timing and peak prediction is very good for each of them and there is little to differentiate their performance.

Although the instance-based and M5 model trees were only marginally worse than the neural network, they have the following advantages:

- speed and ease of development: these models were generated in approximately 2 minutes on a Pentium 1.7 GHz machine, and the data require less pre-processing than a neural network.
- no generalisation techniques like stop training required, which means there is less trial-and-error, one of the main criticisms of the use of neural networks.

These development features may help promote the uptake of this particular technique in an area where there is still skepticism surrounding the use of data-driven models.

Another theoretical advantage of M5 model trees is their transparency. Normally they create a highly interpretable set of rules with regression equations at the leaves of the tree. For example, Khan and See (2006) produced M5 model trees for the River Ouse for 6 and 24 hour ahead forecasting. The 6 hour ahead model produced four rules while the 24 hour ahead forecasting model consisted of a single regression equation that indicated the importance of a single upstream station at this lead time. See et al. (2006) used M5 model trees to integrate different conceptual and AI-based rainfall-runoff models for a catchment in China. The model consisted of two rules which clearly showed the importance of the different individual models under certain flow regimes. However, in this situation the model produced 38 rules, even with pruning of the model tree. Although the model itself produces good results, it was not possible to examine the behaviour of the model, which is one of the main selling points of this data-driven approach. The number of rules produced will be a function of how complex the modeling problem is and may therefore not always provide interpretable rules.

7 Conclusions

In this chapter different data fusion techniques were presented for integrating individual hydrological models that had been developed for the River Ouse catchment. Compared to the individual models, the data fusion approaches almost always performed better in terms of global goodness-of-fit statistics. In addition to performance, the advantages of data fusion approaches are the ability to take evidence from several sources and converge them into a single better informed result, as well as ease of development, especially with approaches such as instance-based learning and M5 model trees. Disadvantages include the need to develop more models for the same forecasting problem.

This is relatively easy when it comes to data-driven modeling but this may be more time consuming when considering physical and conceptual models. However, where these models already exist, these types of methods should be viewed as complementary. All too often real-time forecasting is undertaken using one physically-based model. Data fusion and other ensemble techniques have much to offer operational forecasting yet there is still a reluctance to use these methods due to their black box nature. Although not demonstrated in this particular example, M5 model trees do have the added advantage of transparency, normally producing a small number of comprehensible and interpretable rules. Providing greater understanding as to when different models work best is something that might provide decision makers with a greater confidence in their operational models. Moreover, higher level data fusion architectures could provide a valuable enhancement to operational and real-time forecasting by allowing expert knowledge to be incorporated into the management of water resource systems in the future.

References

Abbott MB, Bathurst JC, Cunge JA, O'Connell PE, Rasmussen J (1986) An introduction to the European Hydrological System – Système Hydrologique Européen, "SHE", 2: Structure of physically-based, distributed modelling system. Journal of Hydrology 87: 61–77

Abrahart RJ, See LM (2002) Multi-model data fusion for river flow forecasting: an evaluation of six alternative methods based on two contrasting catchments. Hydrology and Earth System Sciences 6: 655–670

Ajami NK, Duan QY, Gao XG, Sorooshian S (2006) Multimodel combination techniques for analysis of hydrological simulations: Application to distributed model intercomparison project results. J. Hydrometeorol. 7(4): 755–768

Anctil F, Lauzon N (2004) Generalisation for neural networks through data sampling and training procedures, with applications to streamflow predictions. Hydrology and Earth System Sciences 8: 940–958

Babovic V, Keijzer M (2002) Rainfall runoff modelling based on genetic programming. Nordic Hydrology 33(5): 331–346

Breiman L (1996) Bagging predictors. Machine Learning 26: 123–140

Brion GM, Neelakantan TR, Lingireddy S (2001) Using neural networks to predict peak Cryptosporidium concentrations. Journal of the American Water Works Association 93: 99–105

Box GEP, Jenkins GM (1976) Time Series Analysis: Forecasting and Control. Holden-Day: Oakland

Chaves P, Chang F-J (2006) Simulation of multiple water quality parameters using artificial neural networks. In: Proceedings of the 7th International Conference on Hydroinformatics, September 4–8 2006, Nice, pp 749–756

Chaves P, Kojiri T, Yamashiki Y (2003) Optimization of storage reservoir considering water quantity and quality. Hydrol. Processes 17: 2769–2793

Clair TA, Ehrman JM (1998) Using neural networks to assess the influence of changing seasonal climates in modifying discharge, dissolved organic carbon, and nitrogen export in eastern Canadian rivers. Water Resources Research 34(3): 447–455

Dasarathy BV (1997) Sensor fusion potential exploitation – Innovative architectures and illustrative applications. Proceedings of the IEEE 85: 24–38

Dawson CW, Abrahart RJ, See LM (2007) HydroTest: a web-based toolbox of statistical measures for the standardised assessment of hydrological forecasts. Environmental Modelling and Software 27: 1034–1052

di Pierro F, Khu ST, Savic D (2006) From single-objective to multiple-objective multiple-rainfall events automatic calibration of urban storm water runoff models using genetic algorithms. Water Science and Technology 54 (6–7): 57–64

Gangopadyay S, Gautam TR, Gupta AD (1999) Subsurface Characterization Using Artificial Neural Network and GIS. J. Comp. in Civ. Engrg. 13: 153–161

Georgakakos KP, Seo DJ, Gupta H, Schaake J, Butts MB (2004) Towards the characterization of streamflow simulation uncertainty through multimodel ensembles. J. Hydrol. 298(1–4): 222–241

Giustolisi O, Simeone V (2006) Optimal design of artificial neural networks by a multi-objective strategy: groundwater level predictions. Hydrological Sciences Journal 51(3): 502–523

Gumrah F, Oz B, Guler B, Evin S (2000) The application of artificial neural networks for the prediction of water quality of polluted aquifer. Water Air and Soil Pollution 119: 275–294

Ha SR, Park SY, Park DH (2003) Estimation of urban runoff and water quality using remote sensing and artificial intelligence. Water Science and Technology 47: 319–325

Hall DL (1992) Mathematical techniques in multisensor data fusion. Artech House: Boston

Hall DL, Linas J (1997) An introduction to multisensor data fusion. Proceedings of the IEEE 85 (1): 6–23

Hall DL, Linas J (2001) Handbook of multisensor data fusion. CRC Press: Boca Raton FL

Haykin S (1999) Neural networks: a comprehensive foundation. McMillan: New York

Heppenstall AJ, See LM, Abrahart RJ (2006) Advances in Neural Network Hydrological Modelling: An Adaptive Co-evolutionary Approach. In: Proceedings of the 7th International Conference on Hydroinformatics, September 4–8 2006, Nice, pp 966–973

Jain A, Srinivasulu S (2004) Development of effective and efficient rainfall-runoff models using integration of deterministic, real-coded genetic algorithms and artificial neural network techniques. Water Resources Research 40, art. no. W04302

Jang J-SR, Sun C-T, Mizutani E (1997) Neuro-fuzzy and soft computing. Prentice Hall: NJ

Jarvie HP, Neal C, Robson AJ (1997) The geography of the Humber catchment. The Science of the Total Environment 194/195: 87–99

Karlsson M, Yakowitz S (1987) Nearest neighbour methods for non-parametric rainfall runoff forecasting. Water Resour. Res. 23: 1300–1308

Keijzer M, Babovic V (2000) Genetic programming, ensemble methods and the bias/variance tradeoff – Introductory investigations. Lecture Notes in Computer Science 1802: 76–90

Khan SA, See LM (2006) Rainfall-runoff modelling using data driven and statistical methods, In: Proceedings of ICAST 2006, Islamabad, Pakistan, 2–3 September 2006, IEEE Catalog number: 06X1482, ISBN for CD-Rom: 1-4244-0515-7, ISBN for printed version: 1-4244-0514-9

Khondker MUH, Wilson G, Klinting A (1998) Application of neural networks in real time flash flood forecasting. In: Babovic V, Larsen CL (eds) Proceedings of the Third International Conference on HydroInformatics. AA Balkema: Rotterdam, pp 777–782

Kitanidis PK, Bras RL (1980) Real-time forecasting with a conceptual hydrologic model. 2. Application and results. Water Resour. Res. 16(6): 1034–1044

Kohonen T (1984) Self-organization and associative memory. Springer-Verlag: Berlin

Kojiri T (1999) Cluster-based hydrological analysis and prediction of long-term climate change and drought discharge. Hydrological Extremes: Understanding, Prediction Mitigating, IAHS Publ. No.255, 45–54

Kojiri T, Hanatani S (2002) Knowledge-based decision support system for reservoir operation with knowledge acquisition process on flood prediction and risk accuracy. Flood Defence '2002, Vol.II, Beijing, 1430–1437

Kosko B (1992) Neural networks and fuzzy systems: a dynamical systems approach to machine intelligence. Prentice Hall Inc.: NJ

Kuchment LS, Demidov VN, Naden PS, Cooper DM, Broadhurst P (1996) Rainfall-runoff modelling of the Ouse basin, North Yorkshire: an application of a physically-based distributed model. Journal of Hydrology 181: 323–342

Lane S (2002) More floods, less rain: changing hydrology in a Yorkshire context. The Yorkshire and Humber Regional Review 11: 18–19

Law M, Wass P, Grimshaw D (1997) The hydrology of the Humber catchment. The Science of the Total Environment 194/195: 119–128

Lindsay JB, Shang JQ, Rowe RK (2002) Using complex permittivity and artificial neural networks for contaminant prediction. J. Envir. Engrg. 128: 740–747

Longfield S (1998) River response to recent environmental change in the Yorkshire Ouse basin, northern England. Ph.D. thesis, University of Leeds

Maier HR, Dandy G (1996) The use of artificial neural networks for the prediction of water quality parameters. Water Resources Research 32: 1013–1022

Marshall L, Sharma A, Nott D (2006) Modeling the catchment via mixtures: Issues of model specification and validation. Water Resour. Res. 42(11): W07410

Minns AW, Hall MJ (1996) Artificial neural networks as rainfall-runoff models. Hydrological Sciences Journal 41: 399–417

Nayak PC, Sudheer KP, Rangan DM, Ramasastri KS (2004) A neuro-fuzzy computing technique for modeling hydrological time series. Journal of Hydrology 291: 52–66

Oudin L, Andreassian V, Mathevet T, Perrin C, Michel C (2006) Dynamic averaging of rainfall-runoff model simulations from complementary model parameterizations. Water Resour. Res. 42(7): W07410

Quinlan JR (1992) Learning with continuous classes. In: Proceedings of AI'92, 5th Australian Joint Conference on Artificial Intelligence, World Scientific, Singapore, pp 343–348

Salas JD, Markus M, Tokar AS (2000) Streamflow forecasting based on artificial neural networks. In: Govindaraju RS, Ramachandra Rao A (eds) Artificial Neural Networks in Hydrology, Kluwer Academic Publishers, Dordrecht, pp 23–51

Schmitz GH, Schutze N, Petersohn U (2002) New strategy for optimizing water application under trickle irrigation. J. Irrig. and Drain. Engrg. 128: 287–297

See LM, Openshaw S (1999) Applying soft computing approaches to river level forecasting. Hydrological Sciences Journal 44: 763–778

See LM, Openshaw S (2000) A hybrid multi-model approach to river level forecasting. Hydrological Sciences Journal 45: 523–536

See LM, Shamseldin AY, Abrahart RJ, Dawson CW, Heppenstall AJ (2006) Rainfall-runoff modelling using a committee approach, EGU, Vienna, Austria, 5–9 Apr 2006

Shamseldin AY (1997) Application of a neural network to rainfall-runoff modelling. Journal of Hydrology 199: 272–294

Shamseldin AY, O'Connor KM (1996) A nearest neighbour linear perturbation model for river flow forecasting. J. of Hydrology 179: 353–375

Shamseldin AY, Nasr AE, O'Connor KM (2002) Comparison of different forms of the multi-layer feed-forward neural network method used for river flow forecasting. Hydrology and Earth System Sciences 6: 671–684

Singh RM, Datta B, Jain A (2003) Identification of unknown groundwater pollution sources using artificial neural networks. J. Water Resour. Plng. and Mgmt 130: 506–514

Smith J, Eli RN (1995) Neural network models of rainfall-runoff process. Journal of Water Resources Planning and Management 121: 499–509

Smith D, Singh S (2006) Approaches to multisensor data fusion in target tracking: a survey. IEEE Transactions on Knowledge and Data Engineering 18: 1696–1710

Solomatine DP (2005) Data-driven modelling and computational intelligence methods in hydrology. Encyclopedia of Hydrological Sciences, Wiley

Solomatine DP, Dulal KN (2003) Model tree as an alternative to neural network in rainfall-runoff modelling. Hydrological Sciences Journal 48: 399–411

Solomatine DP, Siek MB (2006) Modular learning models in forecasting natural phenomena. Neural Networks 19: 215–224

Toth E, Brath A, Montanari A (2000) Comparison of short-term rainfall prediction models for real-time flood forecasting. J. of Hydrology 239: 132–147

US Department of Defense (1991) Data fusion subpanel of the joint directors of Laboratories, Technical Panel for C3, Data Fusion Lexicon

Wheater HS, Jakeman AJ, Beven KJ (1993) Progress and directions in rainfall-runoff modelling. In: Modelling Change in Environmental Systems, Jakeman AJ, Beck MB, McAleer MJ (eds), 101–132, Wiley

Xiong LH, Shamseldin AY, O'Connor KM (2001) A non-linear combination of the forecasts of rainfall–runoff models by the first-order Takagi-Sugeno fuzzy system. J. of Hydrology 245: 196–217

Yan SQ, Minsker B (2006) Optimizing groundwater remediation designs under uncertainty using dynamic surrogates and noisy genetic algorithms. In: Proceedings of the 7th International Conference on HydroInformatics, September 4–8 2006, Nice, pp 2513–2520

Yang C-C, Prasher SO, Lacroix R, Sreekanth S, Patni NK, Masse L (1997) Artificial neural network model for subsurface-drained farmlands. J. Irrig. and Drain. Engrg., ASCE 123: 285–292

Young P (2002) Advances in real-time flood forecasting. Philosophical Transactions of the Royal Society London A 360(1796): 1433–1450, DOI: 1410.1098/rsta.2002.1008

Young P (2003) Top-down and data-based mechanistic modelling of rainfall-flow dynamics at the catchment scale. Hydrol. Processes 17(11): 2195–2217

A New Paradigm for Groundwater Modeling

Shu-Guang Li and Qun Liu

Department of Civil and Environmental Engineering, Michigan State University

Summary. A 1987 landmark National Science Foundation report on scientific computing and visualization (McCormick et al., 1987) envisioned the future of scientific computing to be real-time interactive with the modelers being dynamically-engaged and in full control throughout the computational process. The report stressed: scientists not only want to solve equations or analyze data that results from computing, they also want to interpret what is happening to the data during computing. Researchers want to steer calculations in real-time; they want to be able to change assumptions, conceptual framework, resolution, or representation, and immediately see the integrated effects, the ultimate implications, and the complex interrelationships presented intelligently in a meaningful context. They want to be an equal partner with the computer, interact on-line with their data, and drive in real-time the scientific discovery process. While this would certainly be the preferred modus operandi and is finally becoming computationally feasible even for many 3D dynamic problems on a personal computer, it is not the current standard of groundwater modeling. Although these thoughts were first reported nearly twenty years ago, they express an idea that is current and more relevant than ever before as the computing power continues to grow exponentially.

In this chapter, we present a new computing paradigm and a novel, sophisticated computational environment that enables fully taking advantage of today's computing power, especially the computer of the future, and allows, for the first time, real-time 3D groundwater modeling. The new environment, called Interactive Ground Water (IGW), utilizes a powerful "parallel computing", "dynamic visualization", and "computational steering" methodology, restructuring and integrating the entire modeling process. This environment enables seamless, dynamic data routing and fusion of flow modeling, transport modeling, subscale modeling, uncertainty modeling, geostatistical simulation, GIS mapping, and 3D visualization. IGW functions as an intelligent "numerical research laboratory" in which a modeler can freely explore: visually creating aquifers of desired configurations, interactively applying stresses, and then investigating on the fly the geology, dynamic flow and transport in three space dimensions. At any time, a modeler can edit, monitor and interact with virtually any aspects of the integrated modeling process; the modeler can initiate, pause, or resume particle tracking, plume modeling, multi-scale modeling, stochastic modeling, and analyses. IGW dynamically merges geospatial data, modeling inputs

S.-G. Li and Q. Liu: *A New Paradigm for Groundwater Modeling*, Studies in Computational Intelligence (SCI) **79**, 19–41 (2008)
www.springerlink.com

and outputs, and stratigraphic and sampling network information into composite 2D and 3D graphical images – providing a more complete view of the complex interplay among the geology, hydrology, and transport. The capabilities of real-time simulation, analysis, steering, and presentation expand the utility of groundwater modeling as a tool for research, education, professional investigation, and integrated management decision support.

1 Introduction

The enormous increase in computational speed and capacity achieved over the last two decades is responsible for both the development of computational science and engineering and its current status as a unique and powerful tool for scientific discovery. Model-based simulation, a key branch of this new discipline, provides the capability for simulating the behavior of complex systems under realistic environmental conditions. Modeling creates a new window into the natural world [Sack, 1999].

Our understanding of subsurface flow and contaminant transport stands to benefit immensely from model-based research. Models provide the ability to simulate the behavior of integrated, large-scale systems and interactions; they permit prediction of future outcomes based on current or historical conditions. Modeling can provide fundamental insights into complex field-scale behavior of heterogeneous processes, nonlinear scale effects, aquifer system interactions, groundwater and surface water connections, and interactions between geological, hydrological, and biochemical processes. Model-based simulation provides a systematic framework for assimilating and synthesizing field information and prioritizing sampling activities. Modeling becomes particularly useful for addressing "what-if" types of questions, testing hypotheses, assessing data-worth and model uncertainty, and evaluating management, monitoring, and cleanup options. Modeling makes it possible for scientists and engineers to see the unseen, to develop new understanding, and to predict the future [Anderson and Woessner, 1992; Bear, 1979; Bredehoeft, 2002].

1.1 A Substantial Gap

However, practical implementation of groundwater models can be difficult, especially for large-scale, integrated simulation of coupled processes and when available data is limited.

The traditional modeling paradigm employed by most groundwater modelers makes use of a sequential scheme based on disjointed batch simulations and offline visualizations and analyses. A bottleneck occurs under this paradigm because it creates a fragmented modeling process characterized by a significant waste in human efforts and computations, loss of information, and extra disk storage and offline operations. And modelers repeatedly rely on this inefficient process as they refine the conceptual model, through an iterative, trial and error process that is central to the "art" of all modeling. The sequential nature of the modeling scheme significantly increases the turnaround time between questions and final answers. It presents serious obstacles in the process of professional investigation and scientific discovery. It breaks one's train of thought – providing little intuition into the complex interplay among geology, hydrology, chemistry, and management decision making, and little "flow" in the process of subsurface exploration, integrated problem solving, and investigation.

In today's professional, research, and business environment – characterized by high complexity, short development cycle, and high human cost, the ineffective modeling paradigm is turning away many people who may otherwise benefit significantly from model-based simulations. The fragmented paradigm makes people think small by using highly simplified analytical models and creates a substantial gap between what is computationally possible and practically feasible and between basic research and practice [Sack, 1999; Atkins et al., 2002].

1.2 Traditional Modeling Paradigm

Most of today's groundwater modeling studies are interdisciplinary and require simulating coupled processes and solving multiple interrelated models (see Box 1). For example, a typical model-based investigation on contaminant fate and transport at a waste disposal site may consist of a number of computer based stages, including 1) groundwater flow modeling, 2) particle tracking analyses, 3) solute transport modeling, 4) aggregated water and solute mass balance computations and analyses, and 5) overall post-processing and analysis with, e.g., a Geographic Information System (GIS) [Anderson and Woessner, 1992; Pinder, 2002].

The inherently uncertain nature of the subsurface system (not just in aquifer parameter values but, most importantly, in the ways one conceptualizes and parameterizes the aquifer processes, stresses, structure, and boundary conditions) requires iterative implementation among the different modeling stages or throughout the "life cycle" of the modeling project [Bredehoeft, 2003; Konikow and Bredehoeft, 1992; Hassan, 2004; Kovar and Hrkal, 2003; Bear et al., 1992]. Each of the modeling stages may in itself be another iterative procedure that consists of the following sequential steps:

1. Create or modify a conceptual model;
2. Assign or modify model stresses, properties, and starting/initial conditions;
3. Solve the governing equations over the entire specified time span and store the results on a disk;
4. Analyze the results using a post-processing/visualization package [e.g., GIS];
5. Compare with field data;

6. Make appropriate changes to the model;
7. Repeat.

The need to iterate both within and among the disconnected sequential stages makes modeling an inefficient process. It makes it difficult to investigate the conceptual uncertainty and to calibrate the integrative modeling system. Under the traditional sequential paradigm, modelers typically go offline to change the conceptual model or the computational scheme via a graphical interface (e.g., the powerful Department of Defense Groundwater Modeling System-GMS), and each change in the model or the input parameters causes most of the other steps or the various modeling stages in the process to be repeated. Errors as simple as an incorrect value of a single model parameter may only become apparent, in many cases, after many man-hours or a long "calendar time" because of the fragmented nature of the modeling process and the difficulty imposed by visualizing and analyzing results offline during post-processing. For this reason, even major errors invalidating the results of an entire simulation may go undetected. A meticulously calibrated model at one particular stage (e.g., a flow model) may be invalidated at any of the following stages (e.g., during transport modeling, subscale modeling, mass budget calculations, analyses or visualization of the overall results) – forcing the whole inefficient modeling processes within and among the different stages to start over, perhaps again and again!

Model 1 (e.g., flow)

t_1 – Incremental computation
t_2 – Incremental computation
t_3 – Incremental computation
...
t_n – Incremental computation
End
save results to disk or memory
offline postprocessing
offline analysis (e.g., mass balance)
offline visualization

Model 2 (e.g., particle tracking)

t_1 – Incremental computation
t_2 – Incremental computation
t_3 – Incremental computation
...
t_n – Incremental computation
End

save results to disk or memory
offline postprocessing
offline analysis
offline visualization

Model 3 (e.g., solute transport)

t_1 – Incremental computation
t_2 – Incremental computation
t_3 – Incremental computation
...
t_n – Incremental computation
End
save results to disk or memory
offline postprocessing
offline analysis (e.g., mass balance)
offline visualization

...

Integrated offline postprocessing, analysis, mapping, and visualization

Box 1 – The traditional sequential modeling paradigm. The fragmentation in the overall modeling process makes it difficult to take full advantage of the dramatically increased computing power.

We believe the traditional disjointed paradigm is not best suited for integrated modeling that involves multiple steps and stages and requires iterative conceptualizations. We have always known the critical importance of conceptual modeling and that solving groundwater problems is inherently a process of iterative hypothesis testing and involves a significant element of "art" that requires frequent human interactions [Bredehoeft, 2003; Bear et al., 1992; Mercer, 1991; Anderson and Woessner, 1992; Konikow and Bredehoeft, 1992]. But we just do not have an adequate tool that allows exploring potentially large numbers of combinations of possibilities for integrated simulations and to incorporate conceptual changes throughout a multi-staged modeling process. Conceptual modeling typically stops until resources are exhausted, or when the modeler "gives up" when he/she no longer has the time to carry out another model run, even though major questions are still unresolved or data unexplained.

1.3 A Few Basic Questions

It is time we "step back" and ask ourselves a few basic questions:

- Given the computing power today and especially that of the future, can we make groundwater modeling truly a transparent process of "continuous" integrated problem solving, hypothesis testing, and exploration?
- Does the groundwater modeling process have to be this fragmented?
- Why do we have to divide the modeling of concurrently occurring processes into sequential and isolated computational stages?
- Why do we have to decouple the modeling, post-processing, visualization, analyses, and overall presentation?
- Why do we have to investigate a model to "death" before moving on to the next model when we know in all likelihood we will come back to revise it in response to the feedback obtained from the next stage?
- Why do we have to wait for days, months, or even years before we can see the connections between a simple change in the input or a preliminary assumption and its integrated economic or policy implications when it only takes, for many practical problems, a net computer time of minutes or even seconds (per simulation time step) to compute all the state variables of direct interest?
- How can we deal with the bewildering amount of information associated with integrated, large-scale modeling? How can we route the large volumes of dynamic data streams efficiently across the different modeling steps and disjointed stages? How can we make sense of the large amount of information and present it in ways that best reveal the hidden structure and complex interrelationships in a timely fashion?

1.4 The 21st Century Demands a New Paradigm

These questions clearly influence importantly our practical ability to model, to investigate, and to discover, but they have so far received little attention in the hydrologic and environmental modeling community.

In the general context of scientific computing, Sack [1999] emphasized that "the development of traditional codes has probably reached a limit. The future of modeling lays in research to move to a true distribution of, not only equation solving, but also model building, simulation, information and data management, and visualization interpretation". Sack further stressed that there is an "urgent research need for an open, integrative approach to modeling".

Atkins et al. [2002] described a vision for the future of supercomputing that emphasizes removing the human bottleneck and improving the complete "life cycle" of the modeling process. They stressed the need to use computers intelligently as complete tools, environments, or collaboratories, not just as raw technologies for processing data. The NSF Blue Ribbon report [Branscom et al., 1993] listed "removing barriers to the rapid evolution of high performance computing capability" as a major challenge.

Beazley and Lomdahl [1996] pointed out, in the context of molecular dynamic modeling, that large-scale modeling "is more than just a simulation problem, an analysis problem, a visualization problem, or a user interface problem. It is a combination of all of these things -and the best solution will be achieved when these elements have been combined in a balanced manner". Beazley and Lomdahl also stressed that the underlying source of difficulty in solving large problems is "not the result of having large data sets, a lack of tools, or lack of high performance machines, but the entire methodology of the process".

McCormick et al. [1987] and De Fanti et al. [1987] envisioned the future of scientific computing to be real-time interactive with the modelers being in full control throughout the computational process and being an equal partner with the computer. In particular, they stressed: scientists and engineers not only want to solve equations or analyze data that results from supercomputing; they also want to interpret what is happening to the data during supercomputing. Researchers want to steer calculations in close-to-real-time; they want to be able to change assumptions, conceptual framework, resolution, or representation, and immediately see the integrated effects and the complex interrelationships presented in a meaningful context. They want to interact on-line with their data and drive in real-time the scientific discovery process.

While this would certainly be the preferred modus operandi for most scientists and engineers and is finally becoming computationally feasible even on a personal computer for many problems we are solving today, it is not the current standard of groundwater modeling. Although these thoughts were first reported nearly twenty years ago, they express an idea that is current and in fact more relevant than ever before as the computing power continues to grow exponentially.

2 A Visual Steering Environment for Integrated Groundwater Modeling

In this chapter, we present a new paradigm and a novel, sophisticated computational environment for integrated groundwater modeling – one that promises to eliminate the current bottlenecks and allows truly capitalizing on the rapidly increasing computing power. The new environment, called Interactive Ground Water (IGW), utilizes a powerful "parallel computing" methodology as well as novel, emerging computing concepts, including incremental computing, discrete simulation formalism [Aiello et al., 1998; Bisgambiglia, 2002; Delhow, 1995; Cellier, 1996ab; Zeigler, 1990, 2000], dynamic visualization, and computational steering [Papadopoulos et al., 1998; Parker and Johnson, 1995; Folino and Spezzano, 1999; Eisenhauer, et al., 1994; Surles et al., 1994; Sun, 1997]. The term "parallel computing" used in this chapter does not mean modeling on massively parallel processors but, rather, a new

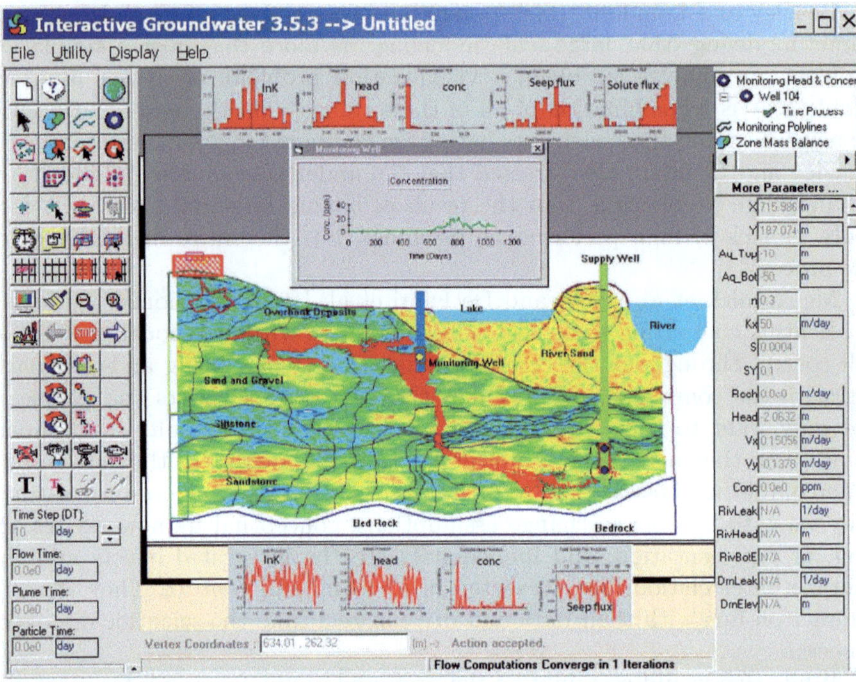

Fig. 1. The IGW environment and a snapshot of an integrated visual simulation and analysis. IGW functions as a "digital groundwater laboratory" in which a user can freely and naturally explore and experiment: creating visually an aquifer of desired configurations and characteristics, interactively applying desired stresses, and then investigating and visualizing on the fly the geology and the dynamic processes of flow and contaminant transport and transformation. The palette on the left provides the tools and buttons for creating, running, visualizing, analyzing, and steering a model. The right pane displays the aquifer conditions and parameter values at the cursor location. (Source: Li and Liu, 2003)

way of structuring computation – one that allows seamless data routing and dynamic integration of groundwater flow modeling, solute transport modeling, data processing, analyses, mapping, and visualization. Figure 1 presents a snapshot of the IGW environment and an illustrative integrated visual simulation and analysis. Some of the basic IGW capabilities were briefly introduced in Li and Liu [2006ab]. Use of IGW as a virtual research laboratory in the classroom was reported in Li and Liu [2003].

A single object-oriented application program forms the core of the new IGW modeling environment. It employs multiple groundwater modeling tasks to rapidly integrate and present complex data in a sophisticated graphical format. It permits the modeler to produce sophisticated 2-D and 3-D graphical displays of spatial, time-varying information at any point during the modeling process. It permits the modeler to steer the entire modeling process.

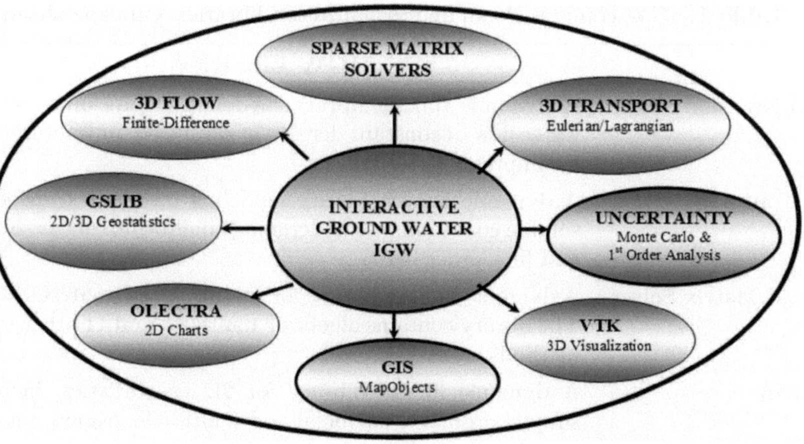

Fig. 2. The IGW computational and visualization engines. The arrows indicate the control capabilities of the various components. The libraries are dynamically embedded in the IGW environment.

Modeling under the new paradigm continually provides and displays results that have been intelligently processed, organized, overlaid, and displayed. It seamlessly and dynamically merges heterogeneous, geo-referenced spatial data into graphical images – integrating related data to provide a more complete view of complex interrelationships. It provides a quick connection between modeling concepts/assumptions and their significance/implications.

We have developed this general-purpose, integrated groundwater modeling environment for research, educational, professional, and outreach pursuits by taking advantage of the recent developments in software engineering, image processing, 3D visualization, geographic information system technologies, as well as, recent research in geostatistics, stochastic subsurface hydrology and modeling, hierarchical multi-scale modeling [Li et al., 2006], and computational methods in subsurface flow and transport modeling.

Figure 2 and Table 1 presents the underlying computational, visualization, and analysis engines dynamically embedded in the IGW environment.

The current version of the 3D FLOW library is limited to modeling homogeneous fluids under isothermal condition in saturated, non-deformable porous media. The 3D transport library is limited to modeling soluble contaminants undergoing simple chemistry such as first-order degradation and equilibrium sorption with a linear isotherm.

3 The New Modeling Paradigm

The new "parallel" paradigm allows us to couple all the various models and solve them for conditions one time step forward from the current time. This

Table 1. IGW computational and visualization libraries and capabilities.

	IGW Libraries
3D Flow	A dynamic linking library modeling steady and unsteady water flow of constant density in saturated, unfractured, and incompressible media.
3D Transport	A dynamic linking library modeling unsteady transport of soluble contaminants undergoing equilibrium linear sorption and first order decay.
Sparse Matrix Solver	A dynamic linking library for solving sparse matrix systems. The library contains algebraic multigrid and 11 other matrix methods.
GSLIB	A dynamic linking library for 2D geostatistics, including simple, ordinary, universal and multiscale kriging, unconditional and conditional simulation using LU decomposition, spectral method, turning bands, sequential Gaussian, simulated annealing.
OLECTRA	A dynamic linking library for plotting hydrograph and concentration breakthrough curves, head and concentration contours, mass balance bar charts, and 3D elevation surfaces.
VTK	A dynamic linking library for integrated visualization of 3D aquifer elevation surfaces, 3D aquifer volumes with local cutout of arbitrary configurations, 3D head and concentration isosurfaces, 2D and 3D head contours, 3D velocity vectors, 3D scatter measurement points, 3D wells and piezometers, 3D aquifer fence diagrams, and basemaps draped over a 3D aquifer elevation surface.
GIS Map-Object	A dynamic linking library for importing GIS shape files (points, polylines and polygons) as base maps or modeling features and automatically extracting the associated modeling attributes (e.g., aquifer elevations, water levels, pumping rates, conductivities, river/lake levels, etc.). The automatic attribute extraction capability is currently limited to the Michigan statewide groundwater database (GIS files).

allows us to restructure and integrate the computations and modeling tasks into a single application program – a program that permits the modeler to visualize the model system's behavior at every time step and evaluate its adequacy, so that we can interrupt the computations, alter the model in significant ways, and restart computations as we deem necessary.

The basic concept is simple. Instead of treating flow and transport separately, we model them concurrently. Instead of treating groundwater flow modeling, solute transport modeling, subscale modeling, particle tracking analysis, and zone budget analyses as different phases in a long sequential process, we couple the multi-staged processes and model them simultaneously. Instead of relegating the graphical presentation of results and their analysis to the "post-processing" phase, at the end of a time consuming sequence of many steps, we incorporate them into a single on-line program along with the simulation, to permit the interpretation of results as soon as they become available at the end of each time step. To accomplish this, we adopt the following new modeling paradigm:

At a discrete time level $t = t_n$ (the nth time step)

1. Flow modeling;
2. Subscale flow modeling, if subarea(s) of detailed interest are specified;
3. Particle tracking, if particles are introduced;
4. Plume transport modeling, if contaminant(s) are introduced;
5. Subscale transport modeling, if subarea(s) of detailed interest are specified and contaminant(s) are introduced;
6. Data and output processing and analysis, solute mass balance and water budget analyses; and
7. Visualization and integrated presentation.
8. $t_n = t_n +$ time step
9. Repeat step 1 to 8.

Box 2 presents a summary of the new "parallel" paradigm under general conditions for integrated groundwater modeling and analysis.

t_1 model 1, model 2, model 3, . . . , on-line processing, analysis, integrated mapping and visualization
t_2 model 1, model 2, model 3, . . . , on-line processing, analysis, integrated mapping and visualization
t_3 model 1, model 2, model 3, . . . , on-line processing, analysis, integrated mapping and visualization
t_4 model 1, model 2, model 3, . . . , on-line processing, analysis, integrated mapping and visualization
. . . .
t_n model 1, model 2, model 3, . . . , on-line processing, analysis, integrated mapping and visualization
end

Box 2 – A new "parallel" modeling paradigm. Data are dynamically routed and IGW goes through the complete modeling process before moving on to the next time step.

The IGW program provides an interactive, graphical environment for defining the aquifer framework, for inputting parameters, properties and stresses, for changing grid resolution, solvers, numerical schemes, and modeling methods, for controlling and managing program execution, and for integrating, overlaying and visualizing data and results.

We have taken advantage of interactive, object-oriented programming and designed the IGW environment so that scientists and engineers can, at any time (including during simulation or analysis), pause to edit and interact on-line with virtually any aspects of the modeling process, (just like what a modeler can do offline at the beginning of the simulation). At any time, the modeler can initiate, stop, and edit particle tracking, plume modeling, sub-scale modeling, and stochastic modeling. At any time, scientists and engineers can see the current results presented in a meaningful way, no matter how preliminary the model inputs or assumptions. The results displayed on the screen can then be used as starting conditions for continued incremental improvement. An incremental modeling capability proves to be very useful for groundwater modeling because of its inherently uncertain nature and the high cost of data acquisition.

Groundwater modeling within the IGW environment becomes a process of high-level graphical conceptualization, as if one is drawing a picture of the site, and iteratively analyzing and improving the mathematical representation of its various features. It becomes a process of iteratively making sense of the results and solving integrated problems. By pointing and clicking the mouse, the modeler can delineate areas of interest (e.g., the spatial extent of the modeled aquifer, its materials and properties; the spatial coverage of rivers, lakes, and wetlands; wells; hydraulic stresses; and contamination sources) and quickly visualize the integrated dynamics and system interaction. The user is always in control throughout the entire modeling/problem solving process.

Specifically, the IGW environment allows an investigator, at any time during the modeling process, to pause program execution and do any of the following:

- To modify the conceptual model: The modeler can input and edit model boundaries, conceptual assumptions, and aquifer structure, properties, and stresses. These changes can be imposed over any graphically specified areas or 3D volumes, independent of the spatial and temporal discretizations employed. And data describing any aquifer property or spatial parameter at scattered locations throughout the modeled region can be analyzed using advanced regression, interpolation, and geostatistical simulation techniques. The conceptual model can be converted on the fly to an integrated numerical model. Figure 3 presents a snapshot of a hierarchically structured "IGW Model Explorer" through which the user can interact, visualize, and edit the conceptual model and the associated properties and stresses. Figure 4 illustrates GIS-based conceptual modeling within IGW and real-time conversion of GIS objects and attributes to a numerical model.
- To modify the numerical representation: The modeler can select and change numerical parameters such as time step and grid spacing, the number of computational layers in a geological layer, the discretization schemes, solution methods, solver parameters, and spatial interpolation

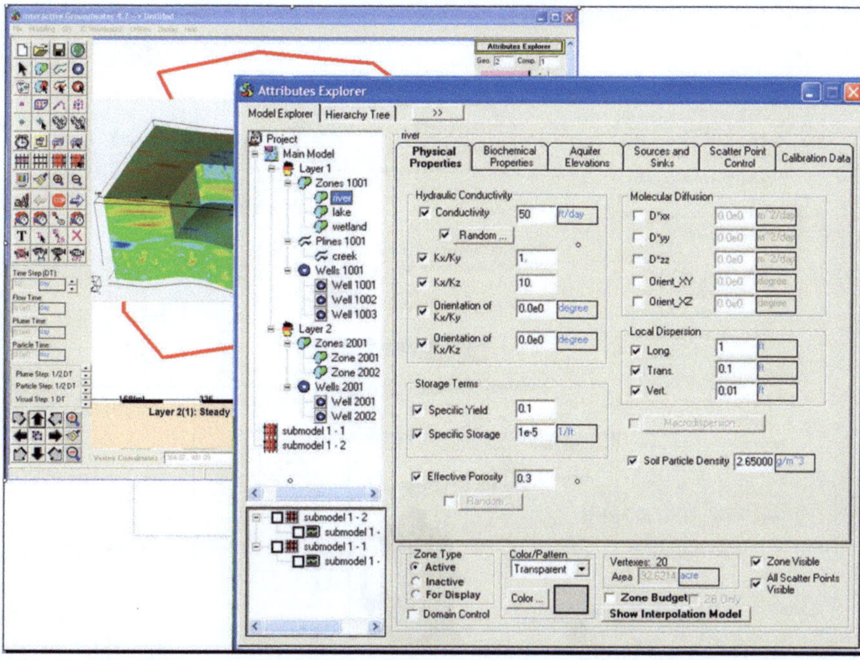

Fig. 3. The IGW model explorer. The model explorer presents a hierarchical visualization of the conceptual model structure and allows a user to navigate across the aquifer system and edit the aquifer features, properties, and stresses over any graphically specified areas or 3D volumes independent of the computational grid. And data describing any aquifer property or spatial parameter at scattered locations throughout the modeled region can be analyzed using advanced regression, interpolation, and statistical simulation techniques. The conceptual model can be converted on the fly to an integrated numerical model.

techniques. Figure 5 shows a snapshot of IGW interfaces through which a user can access and edit on-line the numerical representation and interact with the numerical schemes, the matrix solvers, and the geostatistical interpolation and simulation methods.

- To initiate particle tracking and/or reactive contaminant transport modeling: The modeler can graphically and interactively release particles at a point, along a poly-line, over a polygon, or around the wells and track forward or backward particle migration. The modeler can also simulate the migration of concentration plumes resulting from a number of resources. These include polluted rivers and lakes, polluted rainfall and artificial recharge, waste-well injections, as well as, instantaneous spills and continuous sources with a time-dependent loading rate. Figure 6 presents a typical integrated flow, transport, particle tracking simulation and on the fly visual monitoring and mass balance analyses.

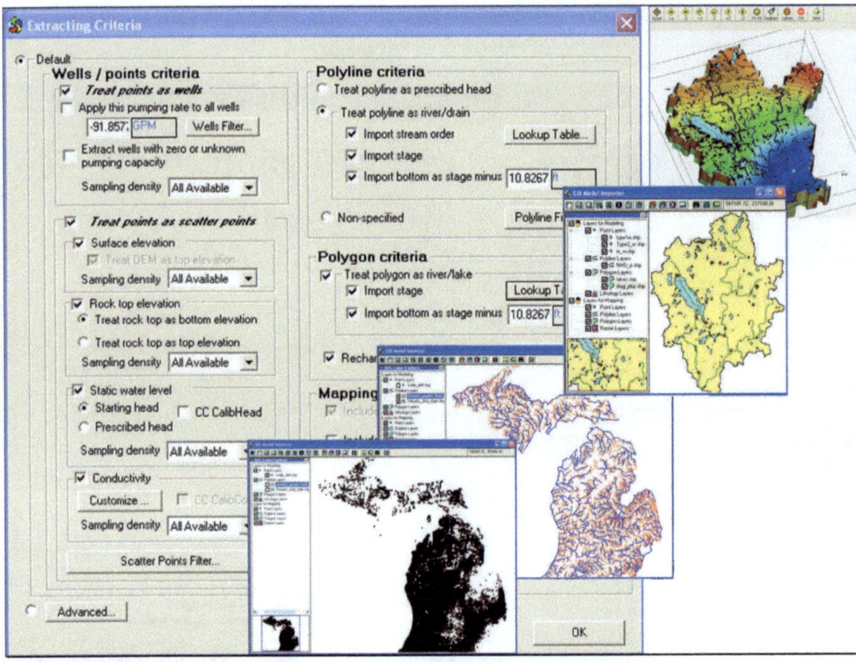

Fig. 4. Real-time GIS-Enabled Modeling. IGW allows quickly building a complex, georeferenced, conceptual model from GIS objects (e.g., streams, lakes, wetlands, wells, digital elevation model, etc.) and attributes (e.g., pumping rates, lake elevations, stream elevations, stream orders, aquifer elevations, lithologic information, etc.) and automatically converting the GIS-based conceptual model into a numerical model. IGW provides a suite of systematic filters controlling the extraction of GIS objects and attributes. IGW is live-linked to Michigan Statewide RS/GIS surface water and groundwater database.

- To develop nested sub-models of flow and transport: The modelers can define incrementally a hierarchy of sub-model regions within a larger model. Sub-models can span one or more geological layers and run in parallel within the parent model. They are solved right after the parent solution is obtained for each time step. Initial and boundary conditions for the sub-models are extracted dynamically and automatically from their parent model at every time step. Figures 7 and 8 present illustrative examples of on the fly, integrated hierarchical modeling of a 2D and 3D groundwater system across multiple spatial scales.
- To examine the impact of unmodeled small-scale heterogeneity, data limitations, and uncertainty: The modeler can perform stochastic first-order analysis [Ni and Li, 2005; Ni and Li, 2006] or Monte Carlo simulations to quantify model uncertainty caused by subgrid, small-scale heterogeneity. IGW Monte Carlo simulation permits any spatial parameters (e.g., conductivity, porosity, partitioning coefficient, decay coefficient, recharge, aquifer

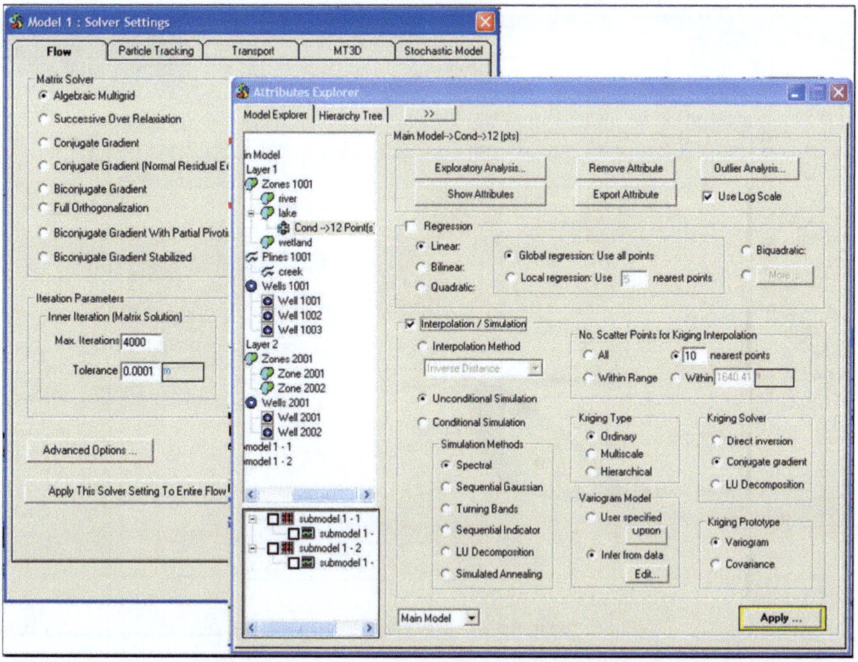

Fig. 5. Dynamic Interaction with the Computational Engines. IGW provides an integrated interface that allows a user to access and edit on-line the numerical representation (e.g., time step, grid spacing, and number of computational layers) and to experiment in real-time with the numerical schemes, matrix solvers, and spatial interpolation methods and quickly see the impact on integrated flow and transport dynamics.

elevations, leakage factor, etc.) to be modeled as a random field, and any temporal stress (e.g., fluctuating surface water elevation, pumping rates, recharge, source concentration and contaminant loading) to be modeled as a 1-D stochastic process. Figure 9 shows an illustrative example of integrated stochastic modeling and on the fly recursive probabilistic analysis.

- To graphically present model characteristics and results and customize the presentation: IGW allows 1) dynamic "data fusion" and integrated presentation of flexible combinations of data, model inputs and outputs, and geo-referenced GIS information for 2D and 3D graphic displays, 2) computing and graphically displaying solute and water fluxes and/or water budgets over any specified zones or along any specified "compliance surfaces", 3) computing and graphically displaying heads and contaminant concentrations as a function of time at monitoring wells. Figure 10 presents illustrative examples of integrated, live-linked, 3D visualization of groundwater flow, solute transport, as well as, the geologic framework, monitoring network, and scattered observational data.

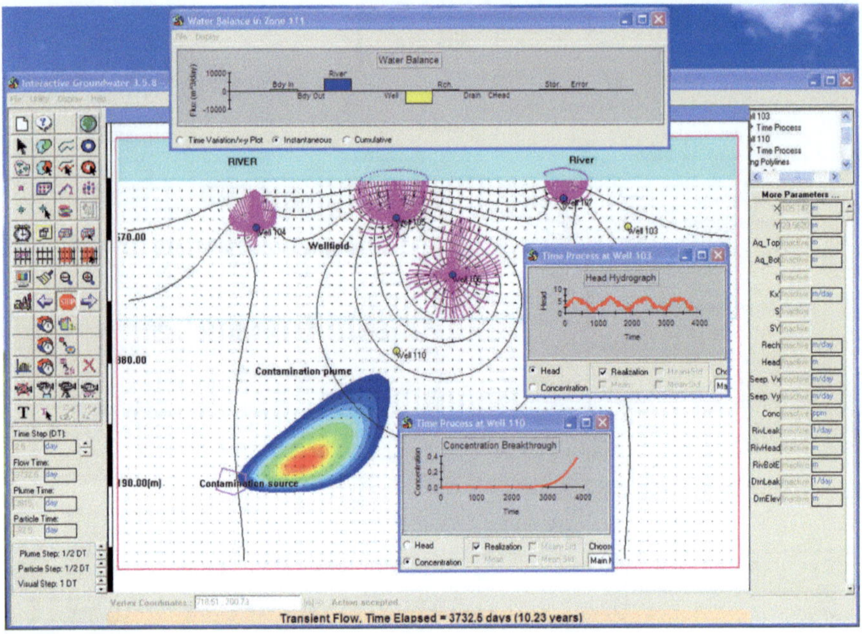

Fig. 6. Real-time Transport Modeling. IGW allows a user to initiate particle tracking and/or reactive contaminant transport modeling anytime during the integrated simulation process. The modeler can graphically and interactively release particles at a point, along a polyline, over a polygon, or around the wells and track forward or backward the particle migration. The modeler can also simulate the migration of concentration plumes originating from a variety of contamination sources. (Source: Li and Liu, 2003)

3.1 Real-time Steering

Rapid, interactive modeling and visualization makes the scientist or engineer an equal partner with the computer in manipulating and maneuvering the 3D visual presentations of the modeling results. It allows the investigator to interactively steer the computation, to control the execution sequence of the program, to guide the evolution of the subsurface flow and plume migration dynamics, to control the visual representation of data during processing, and to dynamically modify the computational process during its execution. Such a sophisticated navigation process would be an invaluable tool for understanding fundamental processes and for practical site investigation.

To maximize the system's flexibility, we have further designed the IGW environment to allow the modeler to adjust the degree of steering at any time, from extremely fine to very coarse. Specifically, the IGW environment is designed such that an investigator can visually step through:

- the "inner iterations" or the iterative process of solving a sparse matrix system which provides an intuitive feel for the rate of iterative convergence

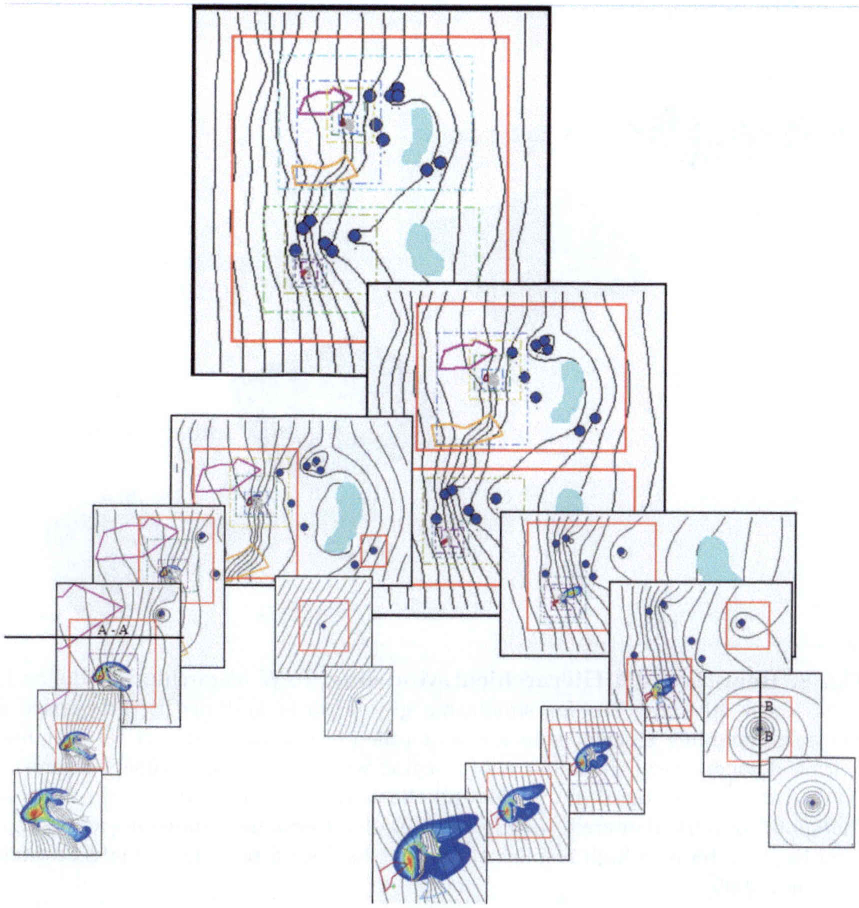

Fig. 7. Real-time Hierarchical Modeling. IGW allows investigating a complex groundwater system across multiple spatial scales. A user can obtain high resolution dynamics in areas of critical interest (e.g., around wells, contamination hotspots) by developing a hierarchy of groundwater models of increasingly higher resolution and smaller domain. IGW automatically couples the model hierarchies, with the parent model dynamically providing the boundary conditions for its "children" which in turn provide the boundary conditions for their own "children". (Source: Li et al., 2006)

and the performance of the matrix solver. In many cases, this pinpoints visually and directly the cause of many commonly encountered numerical problems (e.g., slow convergence or divergence caused by bad inputs, localized singular characteristics, localized extreme heterogeneity, locally very thin geological layer);

- the "outer iterations" or the iterative process of solving the nonlinear governing groundwater equations (e.g., for unconfined aquifers) which is useful

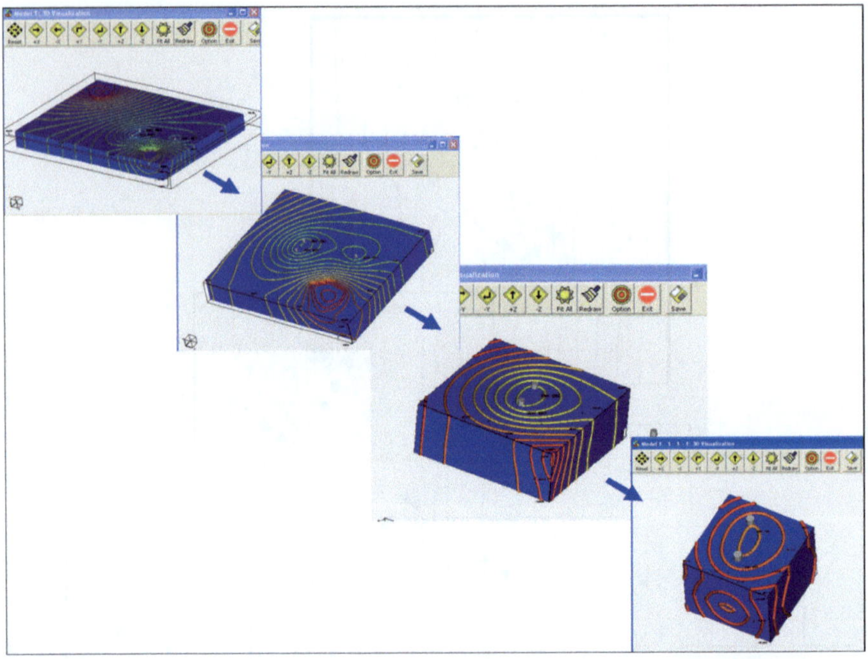

Fig. 8. Real-time 3D Hierarchical Modeling. IGW hierarchical modeling becomes most advantageous for simulating groundwater systems in three-space dimensions, enabling efficient solution of detailed horizontal and vertical dynamics. A typical hierarchical modeling effort begins with a regional, coarsely discretized 2D or a multi-layer quasi-3D model and then progressively "zooms" into localized "hotspots" of critical interest with full 3D details. Hierarchical modeling allows modeling large systems in high resolution without having to solve large matrix systems [Li et al., 2006].

for helping a scientist to obtain an intuitive feel for the nonlinear aquifer dynamics. This also helps pinpoint directly and visually possible sources of common numerical problems associated with nonlinear iterations (e.g., solution divergence or slow convergence caused by highly nonlinear, locally de-saturated aquifer dynamics);

- the hierarchical modeling process which provides an intuitive feel for the connection among flow and transport processes at different spatial scales (e.g., among regional scale, local scale, site-scale, and local hotspots);
- the time increments which is set as the default steering mode. It allows scientists and engineers to visualize "instantly" the aquifer and plume dynamics in a naturally animated fashion. This also provides flexibility and efficiency in the flow and transport simulations and allows cutting adaptively the time-step size when the simulation becomes difficult (e.g., when a plume moves close to a localized heterogeneity or an area in which

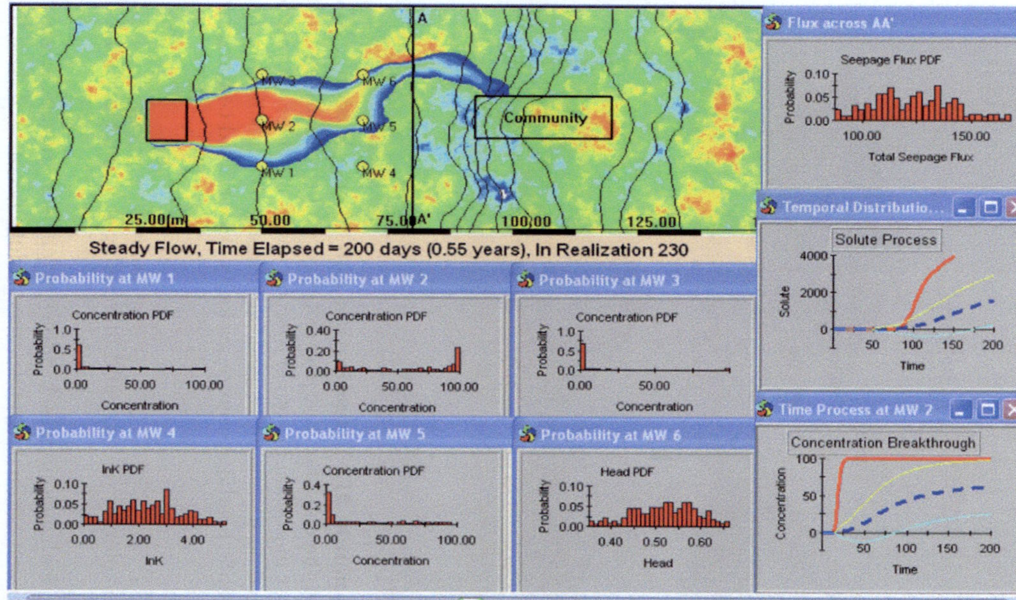

Fig. 9. Real-time Stochastic Modeling. IGW allows a user to examine the impact of unmodeled small-scale heterogeneity through Monte Carlo simulation as well as a perturbation-based first-order uncertainty analysis. IGW Monte Carlo simulation permits any spatial parameters to be modeled as a random field, and any temporal stress to be modeled as a 1-D stochastic process. Flow and transport simulations are automatically "recomputed" for the various property and/or stress realizations. The most recent realizations are employed as they become available to generate point statistics (e.g., probabilities at any interactively specified monitoring well) and spatial statistics (means, uncertainty, and correlations) that can be mapped and visualized as the simulation proceeds. Best available probabilistic characterizations are presented and recursively improved or updated as the number of realizations increases. (Source: Li and Liu, 2003)

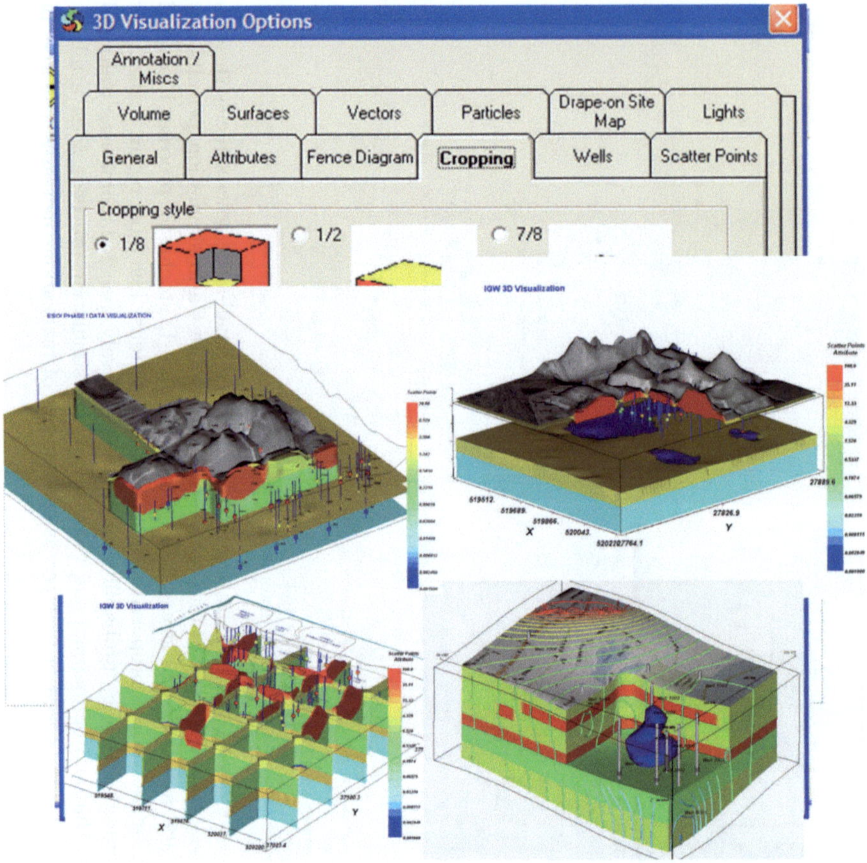

Fig. 10. Integrated, Dynamic 3D Visualization and "Data Fusion". IGW allows a user to investigate the complex interplay between geology, hydrology, flow system, and solute transport migration in aquifer systems. IGW provides automatic, customizable, 3D integrated visualization of geologic framework, monitoring network, scattered data, hydraulic head distribution, velocity vectors, and contaminant plumes.

a sharp change in the velocity occurs) and increasing it when the difficulty passes; and,

- the stochastic model realizations which allows scientists and engineers to visualize how heterogeneity translates into uncertainty because of data limitation and plausible realizations of flow and plume dynamics. The on-line recursive analysis dramatically decreases the extremely long turnaround time in integrated stochastic modeling. An investigator is able to visualize quickly probabilistic characterizations of the groundwater system that is updated with each additional realization. Although it often takes thousands or even tens of thousands of realizations before the final Monte Carlo simulation converges [Li et al., 2004], one can in most cases obtain a fairly good qualitative feel of the general mean behavior of the system dynamics and the spread around the means after just 20 to 30 realizations.

4 Summary and Conclusion

We have focused in this chapter on the overall groundwater-modeling paradigm because we feel the way we model is becoming increasingly incompatible with today's computer technologies and especially those of the future. The new "parallel" paradigm provides seamless data routing and dynamic fusion of flow and transport modeling, visualization, mapping, and analyses, and enables one to truly capitalize on the recent technological revolution.

With 4 GHz desktops available now, 8 GHz microprocessor technology in the labs and faster than 15 GHz technology clearly in sight, actively-visualized subsurface flow dynamics and contaminant hydrogeology incorporating dynamically-linked, intelligently-integrated technology promises potentially significant scientific, economic, and societal benefits. Combined with the hierarchical patch dynamics modeling methodologies [Li et al., 2006], IGW allows, for many problems, real-time modeling and visualization. The new paradigm eliminates the long standing fragmentation in the modeling process and significantly narrows the gap between what is technologically possible and what is practically implementable. Our actual ability to model, to investigate, and to discover can increase in pace with the rapidly advancing computer technologies. We envision that, with the new "parallel" modeling paradigm and the possible realization of a 50 to 500 GHz capability in approximately 5 to 10 years (Moravec, 1998), scientists and engineers may soon be able to model and investigate three-dimensional flow and transport dynamics, complex interactions, and coupled processes with fewer assumptions and in greater details.

Acknowledgements

The integrated research described in this chapter was partially funded by the National Science Foundation under grants EEC-0088137, DUE-0333613, CISE-0430987, and EAR-0523107.

References

1. Afshari, S., Improved Finite Difference Methods for Modeling Two-Dimensional Flow in General Anisotropic Media. Master's thesis, Department of Civil and Environmental Engineering, Michigan State University, 2003.
2. Anderson, M.P. and W.W. Woessner. Applied Groundwater Modeling. Simulation of Flow and Advective Transport. Academic Press, San Diego, California, 1992.

3. Atkins, D.E., K. Droegemeier, S. Feldman, H. Garcia-Molina, M.L. Klein, D.G. Messer-Schmitt, P. Messina. Revolutionizing Science and Engineering Through Cyber Infrastructure: Report of the National Science Foundation Blue-Ribbon Advisory Panel on Cyberinfrastructure, 2002.
4. Aiello et al. (1998). An Object Oriented Environment for Discrete Event Modeling and Simulation, Proceeding of Software Process Simulation Modeling (Silver Falls, Oregon), pp. 126–138.
5. Bear, J. Hydraulics of Groundwater, McGraw-Hill, 1979
6. Bear, J.M.S. Beljin, and R.R. Ross. Fundamentals of Groundwater Modeling, in EPA Groundwater Issue, EPA/540/S-92/005, 11pp., 1992.
7. Beazley, D.M., P.S. Lomdahl. 1996. Lightweight Computational Steering of Very Large Scale Molecular Dynamics Simulations, In SC1996, Pittsburgh, PA, November 17–22, 1996, Conference Proceedings.
8. Bredehoeft, J.D., From Models to Performance Assessment: The conceptualization Problem. Ground Water, Vol. 41, No. 5. September-October 2003.
9. Bredehoeft, J.D., The Water Budget Myth Revisited: Why Hydrogeologists Model. Ground Water. Volume 40 Number 4, July/August 2002
10. Bisgambiglia, P., M. Delhom, J. Santucci, An efficient and evolutionary hierarchical modeling and simulation approach. Source: Systems Analysis Modeling Simulation archive, Volume 42, Issue 2, February 2002
11. Branscomb, L., T. Belytschko, P. Bridenbaugh, T. Chay, J. Dozier, G.S. Grest, E.F. Hayes, B. Honig, N. Lane, W.A. Lester, Jr., G.J. McRae, J.A. Sethian, B. Smith, M. Vernon. From Desktop To Teraflop: Exploiting the U.S. Lead in High Performance Computing, NSF Blue Ribbon Panel on High Performance Computing. August 1993.
12. Cellier, F.E. (1996). Object-Oriented Modeling: Means for Dealing With System Complexity, Proceedings of 15th Benelux Meeting on Systems and Control, Mierlo (The Netherlands), pp. 53–64.
13. Cellier, F.E. (1996). Object-Oriented Modeling: A Tool supporting Flexible Automation, Proc. WAC'96, 2nd World Automation Congress, (Montpellier, France), 107–112.
14. Clyne, J., 1998. Volsh: A tool for real-time interactive visualization. SCDzine, Winter, Vol. 19, No.1.
15. De Fanti et al. Special issues on visualization in scientific computing. Computer Graphics, 21(6), November 1987.
16. Delhom, M., et al. (1995). Modeling and Simulation of Discrete Event Systems, In: IEEE Conference on Systems, Man and Cybernetics (Vancouver, Canada), 5, 4191–4195.
17. Deutsch, C.V. and Journel, A.G. (1998). GSLIB: Geostatistical software library and users guide. Oxford University Press.
18. Eisenhauer, G., Weiming Gu, Karsten Schwan and Niru Mallavarupu, Falcon – Toward Interactive Parallel Programs: The On-line Steering of a Molecular Dynamics Application, In Proceedings of The Third International Symposium on High-Performance Distributed Computing (HPDC-3), San Francisco, August 1994. IEEE Computer Society. An early version of this paper is also available as technical report GIT-CC-94-08, College of Computing, Georgia Institute of Technology, Atlanta, GA 30332-0280.
19. Environmental Systems Research Institute, Inc. MapObjects, Building Applications with MapObjects, ESRI
20. Folino, G. and G. Spezzano, Bioremediation Experiments using Interactive Computational Steering on High Performance Computers. ISI-CNR, DEIS, Universita della Calabria, Italy Fifth European SGI/Cray MPP Workshop. BOLOGNA (Italy)-September 9–10, 1999
21. Hassan, A.E. Validation of Numerical Ground Water Models Used to Guide Decision Making. Ground Water, Volume 42, Number 2. March-April 2004.
22. Johnson, C.R. and S.G. Parker. Applications in computational medicine using SCIRun: A computational steering programming environment. In Supercomputer '95, pages 2–19. Springer-Verlag, 199524.

23. Kovar, K. and Hrkal, Z. Calibration and Reliability in Groundwater Modeling: A Few Steps Closer to Reality. IAHS Publication 277 (published August 2003).
24. Konikow, L. and Bredehoeft, 1992. Groundwater models cannot be validated. Advances in Water Resources, 1992. Vol. 15, No.1, p.75–83.
25. Krabbenhoft, D. and M.P. Anderson, 1986. Use of a groundwater model for hypothesis testing, Ground Water 24(1), 49–55.
26. Li, S.G., Q. Liu, and S. Afshari, "An Object-Oriented Hierarchical Patch Dynamics Paradigm (HPDP) for Groundwater Modeling". Environmental Modeling & Software. 21 (5): 744–749 MAY 2006.
27. Li, S.G. and Q. Liu, "A Real-time, Computational Steering Environment for Integrated Groundwater Modeling". Ground Water 44 (5): 758–763 SEP-OCT 2006.
28. Li, S.G. and Q. Liu, "Interactive Ground Water (IGW)", Environmental Modeling & Software. 21 (3): 417–418 MAR 2006.
29. Li, S.G., H.S. Liao and C.F. Ni. Stochastic Modeling of Complex Nonstationary Groundwater Systems, Advances in Water Resources. 27(11), pp 1087–1104, 18 pages, 2004.
30. Li, S.G., Q. Liu, Interactive Ground Water (IGW): An Innovative Digital Laboratory for Groundwater Education and Research, COMPUTER APPLICATIONS IN ENGINEERING EDUCATION. Vol. 11(4):179 202, 2003.
31. Liao, H., K.J. Paulson, S.G. Li, C. F. Ni. IGW 3 Reference Manual, Department of Civil and Environmental Engineering, Michigan State University, 2003.
32. McCormick, B.H., T.A. DeFanti, and M.D. Brown, eds., Visualization in Scientific Computing, ACM Press, 1987.
33. Mercer, J.W. 1991. Common Mistakes in Model Applications. Proc. ASCE Symposium on Ground Water, Nashville,Tennessee, July 29–August 2, 1991.
34. Moravec, H. 1998. When will computer hardware match the human brain? Journal of Evolution and Technology. Vol. 1, 1998.
35. Papadopoulos, P.M., J.A. Kohl, B.D. Semeraro, "CUMULVS: Extending a Generic Steering and Visualization Middleware for Application Fault-Tolerance," Proceedings of the 31st Hawaii International Conference on System Sciences (HICSS-31), Kona, Hawaii, January 1998.
36. Parker, S.G. and C.R. Johnson. SCIRun: A scientific programming environment for computational steering. In Supercomputing '95. IEEE Press, 1995.
37. Paulson, K.J. and S.G. Li. Interactive Groundwater Users Manual, Department of Civil and Environmental Engineering, Michigan State University, 2002.
38. Sack, R. "Model-Based Simulation", white paper, National Science Foundation, Arlington. 1999.
39. Schroeder, W.J., K. Martin and B. Lorensen. The Visualization Toolkit: An Object-Oriented Approach to 3D Graphics, Prentice Hall PTR, 1998.
40. Sun, M., 1997. Accelerate and improve prospect analysis with actively-visualized geosciences. American Oil and Gas Reporter, Nov. 1997 issue.
41. Surles, M., Richardson, J., Richardson, D. and Brooks, F. (1994). "Sculpting Proteins Interactively: Continual Energy Minimization Embedded in a Graphical Modeling System." Protein Science, 3, 198–210.
42. Ward, D.S., D.R. Buss, J.W. Mercer, and S. Hughes, 1987. A telescopic mesh refinement modeling approach as applied to a hazardous waste site, Water Resources Research, 23(4):603–617.
43. Zeigler, B.P., Object-oriented simulation with hierarchical, modular models: intelligent agents and endomorphic systems, Academic Press Professional, Inc., San Diego, CA, 1990.
44. Zeigler, B.P., Tag Gon Kim, Herbert Praehofer, Theory of Modeling and Simulation, Academic Press, Inc., Orlando, FL, 2000.

Information Fusion using the Kalman Filter based on Karhunen-Loève Decomposition

Zhiming Lu[1], Dongxiao Zhang[2], and Yan Chen[3]

[1] Earth and Environmental Sciences Division, Los Alamos National Laboratory
[2] Department of Civil and Environmental Engineering, and Mork Family Department of Chemical Engineering and Material Sciences, University of Southern California
[3] Mewbourne School of Petroleum and Geological Engineering, University of Oklahoma

1 Introduction

Properties of porous media, such as hydraulic conductivity and porosity, are intrinsically deterministic. However, due to the high cost associated with direct measurements, these properties are usually measured only at a limited number of locations. The number of direct measurements is definitely not enough to infer the true parameter fields. This problem is further complicated by the fact that medium properties exhibit a high degree of spatial heterogeneity. This combination of significant spatial heterogeneity and a relatively small number of direct observations leads to uncertainty in characterizing medium properties, which in turn results in uncertainty in estimating or predicting the corresponding system responses (such as hydraulic head). Fortunately, it is relatively easy to measure the system responses, which can be used to infer medium properties. With newly developed measurement techniques such as remote sensing and *in-situ* permanent sensors, more observations on system responses become available.

Large efforts have been made to take the advantage of all available observations, both the limited number of direct measurements of medium properties and a larger amount of observations of system responses, to obtain better estimates of the primary parameters of the porous media, thereby reducing the uncertainty associated with model predictions. In hydrology, many inverse models have been developed for aquifer characterization [2, 4, 12, 14, 18, 19, 23, 29, 31, 35, 37]. Review and comparison of some of these inverse models can be found in [11, 26, 36, 41]. In these models, the best estimate of a medium property is obtained by minimizing the mismatch between the estimated values and observed ones. However, these models are not capable of incorporating measurements dynamically, which means that,

Z. Lu et al.: *Information Fusion using the Kalman Filter based on Karhunen-Loève Decomposition*, Studies in Computational Intelligence (SCI) **79**, 43–68 (2008)
www.springerlink.com

if new data are available at a time, one has to run the inverse models again from the very beginning.

Recently, a number of sequential inverse methods (history matching in petroleum enginnering) have been proposed, which is basically extended from the optimal control theory of Kalman filter. These methods include Extended Kalman filter (EKF), Ensemble Kalman filter (EnKF), and their variants. The common ground of these methods is their capability of incorporating observations sequentially, which results in a significant reduction of the number of data dealt at any given time. An up-to-date system, consisting of a best estimate and the corresponding uncertainty represented by a state error covariance matrix, is always provided by these sequential methods. The fundamental difference among these methods is the way by which the state error covariance matrix is estimated. The EKF uses the first-order linearization approach and needs to keep track of the whole covariance function, which is computationally expensive for large-scale problems. The EnKF employs the Monte Carlo method, in which the covariance matrix is updated from a small-sized ensemble (a small number of realizations). Some literature shows that for large-scale highly nonlinear problems the EnKF is superior to the EKF in terms of both efficiency and accuracy [6, 7, 17].

The EnKF is conceptually simple, easy to implement, and capable of accounting for different types of models and observation noises, and its computational cost is relatively low compared to other approaches. The method has been used in a large number of applications in various fields such as meteorology, oceanography, hydrology, and reservoir engineering [6, 7, 13, 20, 27, 28]. The size of the ensemble is crucial for the EnKF, because the standard deviation of the sampling errors of the Monte Carlo method converges very slowly at a rate inversely proportional to the square root of the sample size. To reduce the sampling error by a factor of two, the number of Monte Carlo simulations has to be increased by a factor of four. If the observations are not perfect, the sampling error also appears in the ensemble of observations, since the EnKF needs a set of perturbed observations to model the observation noise. In general, a small ensemble induces a large sampling error, which tends to underestimate the state error covariance; a large ensemble leads to computational inefficiency.

Many methods have been proposed to reduce the sampling error associated with the small-sized ensemble in the EnKF. Ensemble square root filter attempts to avoid the perturbed observations by using different Kalman gains to update the ensemble mean and the deviation from the ensemble mean [34]. Anderson and Anderson [1] used a parameter to broaden the ensemble spread. Double Ensemble Kalman filter divides the ensemble into two parts, and uses the Kalman gain calculated from one ensemble to update the other. These methods showed promising results even with relatively small number of ensemble members. On the other hand, these methods are application-dependent and some parameters are difficult to quantify. Extra effort may need to tune these parameters in order to obtain reasonable results.

For high-resolution, large-scale problems, the dimension of the state in general is fairly large. One major problem of various Kalman filter methods comes down to how to efficiently approximate the state error covariance function in each update with a dimension-reduced approach. The EnKF is one type of dimension-reduced approaches since the number of ensemble members is in general smaller than the dimension of the state. There are other attempts that estimate the state error covariance matrix by selecting some principal modes of the state vector, for instance, singular evolutive extended Kalman filter (SEEK) [30], reduced rank square root filter (RRSQRT) [33], and partially orthogonal ensemble Kalman filter (POEnKF) [16].

Recently, Zhang et al. [40] developed a KL-based Kalman filtering scheme (KLKF), which is a type of dimension-reduced Kalman filtering method based on the Karhunen-Loève (KL) expansion of a medium property and orthogonal polynomial decompositions of dependent variables. In the KLKF method, the covariance of the medium property is efficiently approximated by a small set of eigenvalues and eigenfunctions attributed to the mean square convergence of the KL decomposition. Compared to the full covariance matrix, the finite number of modes used to approximate the covariance represents a significant reduction in random dimensions. In this updating procedure, the forward problem is solved with a moment method based on Karhunen-Loève decomposition (KLME), from which the mean and covariance of the state variables can be constructed, when needed. The statistics of both the medium properties and system responses are then updated with the available measurements at the current time using the auto- and cross-covariance obtained from the forward step. They used a synthetic 2-D example to demonstrate the capability of this new method for a stationary conductivity field and compared the results with those from the EnKF method. Their numerical results show that the KLKF method is capable of significantly reducing the required computational resources with satisfactory accuracy, which indicates the potential applicability of this approach to high-resolution, large-scale predictive models.

This chapter is organized as follows: The problem being addressed is described in detail in Section 2. Section 3 provides some basic concepts and formulations of the data assimilation methods. The Karhunen-Loève decomposition for both stationary and nonstationary hydraulic conductivity fields is given in Section 4. The KL-based moment method for solving the first-oder head and head covariance is outlined in Section 5. In Section 6, the KL-based Kalman filtering scheme is introduced, including a detailed description on updating hydraulic conductivity, head field, and eigenvalues of eigenfunctions of the conditioning covariance. The KLKF method is illustrated using a synthetic example in Section 7, with a detailed discussion on the accuracy and efficiency of the method as compared with the EnKF method. The chapter concludes in Section 8 with a short summary and discussion.

2 Statement of the Problem

We consider transient fluid flow in a heterogeneous porous medium satisfying the following governing equation:

$$\nabla \cdot [K_s(\mathbf{x})\nabla h(\mathbf{x},t)] + g(\mathbf{x},t) = S_s \frac{\partial h(\mathbf{x},t)}{\partial t}, \tag{1}$$

subject to initial and boundary conditions:

$$h(\mathbf{x},0) = H_0(\mathbf{x}), \qquad \mathbf{x} \in D, \tag{2}$$

$$h(\mathbf{x},t) = H(\mathbf{x},t), \qquad \mathbf{x} \in \Gamma_D, \tag{3}$$

$$K_s(\mathbf{x})\nabla h(\mathbf{x},t) \cdot \mathbf{n}(\mathbf{x}) = -Q(\mathbf{x},t), \qquad \mathbf{x} \in \Gamma_N, \tag{4}$$

where $K_s(\mathbf{x})$ is the hydraulic conductivity, $h(\mathbf{x},t)$ is the pressure head, $g(\mathbf{x},t)$ is the source/sink term, S_s is the specific storage, $H_0(\mathbf{x})$ is the initial head in the domain D, $H(\mathbf{x},t)$ is the prescribed head on Dirichlet boundary segments Γ_D, $Q(\mathbf{x},t)$ is the prescribed flux across Neumann boundary segments Γ_N, and $\mathbf{n}$ is an outward vector normal to the boundary $\Gamma = \Gamma_D \cup \Gamma_N$. Here the hydraulic conductivity is considered as a spatially-correlated random variable, while specific storage S_s is treated as a deterministic constant because of its relatively small variability. For simplicity, it is assumed that both initial and boundary conditions are deterministic.

Since K_s is a random function, the flow equations become stochastic partial differential equations, which can be solved through several approaches, such as Monte Carlo simulations, the moment-equation approach [38], and the KL-based moment method (KLME) [21, 22, 39]. All Kalman filter methods need to solve the forward problem, i.e., (1)-(4). The EnKF uses the Monte Carlo simulations to obtain the covariance of the state variable (pressure head), while the KLKF uses the KLME to conduct forward modeling. It has been shown that the KLME method is computationally more efficient than the Monte Carlo simulations [22], which makes it possible to develop an efficient algorithm that takes advantages of both the Kalman filter and the KLME methods.

The hydraulic conductivity is assumed to follow a log normal distribution (see a detailed review in [32]), and we work with the log-transformed variable $Y(\mathbf{x})$, given as:

$$Y(\mathbf{x}) = \ln K_s(\mathbf{x}) = \langle Y(\mathbf{x})\rangle + Y'(\mathbf{x}), \tag{5}$$

where $\langle Y(\mathbf{x})\rangle$ is the ensemble mean of $Y(\mathbf{x})$, representing a relatively smooth unbiased estimate of the unknown random function $Y(\mathbf{x})$, and $Y'(\mathbf{x})$ is the zero-mean fluctuation.

Suppose we have N_Y direct measurements of the log hydraulic conductivity $Y_1, Y_2, \ldots, Y_{N_Y}$, taken at locations $\mathbf{x}_1, \mathbf{x}_2, \ldots, \mathbf{x}_{N_Y}$, and N_h pressure head measurements located at $\chi_1, \chi_2, \ldots, \chi_{N_h}$, measured at some time intervals. As mentioned earlier, many inverse models are not capable of dynamically updating parameter fields. In the following sections, we describe an efficient algorithm to characterize medium properties by dynamically incorporating these measurements when they become available. During this data assimilation process, the stochastic differential flow equation is solved forward with time; while the mean and fluctuation of the hydraulic conductivity are modified together with the computed pressure head to honor all the observations at various times.

3 Basic Concepts of Data Assimilation

Data assimilation is a dynamically-updating process, in which estimates of model parameters and state variables are updated by requiring consistency with observations and governing flow equations. With newly developed measurement techniques such as remote sensing and *in-situ* permanent sensors, more observations become available. However, observations from these techniques are usually indirectly related to model parameters, which are of the most interest in hydrological applications. Moreover, measurements may be corrupted by noise from either known or unknown sources. A suitable approach is necessary to reconcile information from multiple sources. The Kalman filter is a widely used sequential data assimilation method for obtaining a least squares estimation of the state vector of the system [8]. It is capable to take into account various types of observations when they become available.

The Kalman filter addresses the general problem of estimating the state vector $\mathbf{S} \in \Re^n$, which represents the state of the system, including model parameters (e.g., hydraulic conductivity) and dependent variables (e.g., hydraulic head). There are two major stages in the Kalman filter procedure, forecast and assimilation [3]. The forecast stage updates the state vector $\mathbf{S}$ and the covariance matrix $\mathbf{P}$:

$$\mathbf{S}^f(i) = \mathbf{\Phi}\,\mathbf{S}^a(i-1) + \mathbf{e_1}(i), \tag{6}$$

$$\mathbf{P}^f(i) = \mathbf{\Phi}\,\mathbf{P}^a(i-1)\mathbf{\Phi}^T + \mathbf{R_1}(i), \tag{7}$$

where i is the time step, superscripts f and a stand for forecast and assimilation stages, respectively, $\mathbf{\Phi}$ is a linear transition matrix that forwards the state vector at time $i-1$ to the current time i, and $\mathbf{e_1}$ represents the process error, which is a random vector with zero mean and covariance $\mathbf{R_1}$. The forecast model will keep on running with time until new observations become available. Once new data are available, the assimilation stage starts, which can be expressed mathematically as

$$\mathbf{G}(i) = \mathbf{P}^f(i)\mathbf{H}^T\left[\mathbf{H}\,\mathbf{P}^f(i)\mathbf{H}^T + \mathbf{R_2}(i)\right]^{-1}, \tag{8}$$

$$\mathbf{d}(i) = \mathbf{H}\,\mathbf{S}^t(i) + \mathbf{e_2}(i), \tag{9}$$

$$\mathbf{S}^a(i) = \mathbf{S}^f(i) + \mathbf{G}(i)\left[\mathbf{d}(i) - \mathbf{H}\,\mathbf{S}^f(i)\right], \tag{10}$$

$$\mathbf{P}^a(i) = \left[\mathbf{I} - \mathbf{G}(i)\mathbf{H}\right]\mathbf{P}^f(i), \tag{11}$$

where $\mathbf{H}$ is a measurement operator that relates the state vector and observations $\mathbf{d}$, superscript T stands for matrix transpose, and $\mathbf{e_2}$ is the measurement error that has a zero mean and error covariance matrix $\mathbf{R_2}$. The first step in the assimilation stage is to compute the Kalman gain $\mathbf{G}(i)$ using (8), and then observations $\mathbf{d}$ are taken using (9). The next step is to compute a posterior estimate of the state vector in (10) by incorporating new observations, and then update a posterior error covariance from (11).

The standard Kalman filtering scheme as described above requires computing and storing the error covariance matrix of state variables, which is computationally expensive for large-scale problems (say, with millions of or even more grid nodes). To alleviate this problem, ensemble Kalman filter (EnKF), a variant of Kalman filter, has been developed [6]. In the EnKF, one keeps track of a limited number of realizations, rather than the covariance matrix as in the standard Kalman filter. At each update, the required subset of the covariance matrix is then obtained from these realizations. Owing to its conceptual simplicity, relative ease in implementation, and the ability to account for possible model noises/errors, the EnKF has been found useful in a large number of applications in various fields such as meteorology, oceanography, hydrology, and reservoir engineering [7, 13, 27, 28]. The computational cost of the EnKF is relatively low (if only a small number of realizations is used) compared to other Kalman filters. The EnKF is essentially a Monte Carlo method, depending on the number of realizations. The goodness of the covariance function approximated from an ensemble strongly depends on the number of realizations used in the ensemble, and the appropriate number of realizations may depend on the nature of the problems and may not be known *a priori*.

One major problem of various existing Kalman filtering methods is how to efficiently compute and store the covariance function in each update. Their substantial requirement on computational resources prevents us from applying them to high-resolution, large-scale simulation problems.

In this chapter, we introduce an efficient, dimension-reduced Kalman filtering scheme based on Karhunen-Loève expansions of the medium property and orthogonal polynomial decompositions of the state variable [40]. The covariance of the medium property is effectively approximated by a small number of eigenvalues and eigenfunctions using the Karhunen-Loève (KL) decomposition. The same number of first-order head fields (modes) are solved and stored in each update. Compared to the full covariance matrix in the EnKF, the finite number of head modes used to approximate it represents a significant reduction in random dimension. The reconstruction of the covariance functions from the KL decomposition of $Y(\mathbf{x})$ and the first-order head modes can be done whenever needed. In each update, the forward problem is solved using an efficient KL-based moment method [39] that gives a set of functions from which the mean and covariance of the state variables can be constructed, when needed. The statistics of both the medium property and the system responses are then efficiently updated with the available measurements at this time using the auto- and cross-covariances obtained from the forward modeling. The KL-based Kalman filter (KLKF) will significantly reduce required computational resources (both computational time and storage) and allows us to efficiently incorporate continuous observations into high-resolution predictive models for flow and transport in large-scale problems.

4 Karhunen-Loève Decomposition

For a stochastic process $Y(\mathbf{x},\omega) = \ln K_s(\mathbf{x},\omega)$, where $\mathbf{x} \in D$ and $\omega \in \Omega$ (a probability space), the covariance function $C_Y(\mathbf{x},\mathbf{y}) = \langle Y'(\mathbf{x},\omega)Y'(\mathbf{y},\omega)\rangle$ is bounded, symmetric, and positive definite, thus it can be decomposed as [9, 10, 39].

$$C_Y(\mathbf{x},\mathbf{y}) = \sum_{m=1}^{\infty} \lambda_m f_m(\mathbf{x}) f_m(\mathbf{y}), \tag{12}$$

where m is the index of modes, λ_m and f_m are eigenvalues and deterministic eigenfunctions, respectively. The eigenvalues and eigenfunctions can be solved from the following Fredholm equation

$$\int_D C_Y(\mathbf{x},\mathbf{y}) f(\mathbf{y}) d\mathbf{y} = \lambda f(\mathbf{x}). \tag{13}$$

The set of eigenfunctions are orthogonal and form a complete set

$$\int_D f_n(\mathbf{x}) f_m(\mathbf{x}) d\mathbf{x} = \delta_{mn}, \tag{14}$$

where δ_{mn} is the Kronecker delta, $\delta_{mn} = 1$ for $m = n$ and 0 otherwise. In addition, it can be shown from (12) that the summation of all eigenvalues equals the total variability of $Y(\mathbf{x})$, $\sum_{n=1}^{\infty} \lambda_n = \int_D \sigma_Y^2(\mathbf{x}) d\mathbf{x}$. One may sort the set of eigenvalues λ_m in a non-increasing order, and the corresponding eigenfunctions then exhibit a decreasing characteristic scale as the index m increases [9, 39]. The stochastic process $Y(\mathbf{x})$ then can be expanded as:

$$Y(\mathbf{x}) = \langle Y(\mathbf{x})\rangle + \sum_{m=1}^{\infty} \xi_m \sqrt{\lambda_m} f_m(\mathbf{x}), \tag{15}$$

where ξ_m are orthogonal random variables, i.e., $\langle \xi_m \rangle = 0$ and $\langle \xi_m \xi_n \rangle = \delta_{mn}$. In the case of normally distributed Y, ξ_m are standard Gaussian random variables. It has been shown that the KL expansion, i.e., (15), is of mean square convergence and may be well approximated with a finite summation. By truncating (15), one in fact ignores some small-scale variability of Y. The number of modes required for accurately approximating $Y'(\mathbf{x})$ depends on the ratio of the correlation length to the dimension of the domain [39]. In many cases, the magnitude of eigenvalues decays very fast, which means that Y can be approximated using a small number of terms.

Note that stationarity of the process $Y(\mathbf{x})$ is not required in the above derivation. Suppose the simulation domain D is partitioned into K non-overlapping subdomains $D = \cup_{k=1}^{K} D_k$ and $D_m \cap D_n = \phi$ for $m \neq n$, where ϕ stands for the null set, and accordingly the log hydraulic conductivity field can be written as

$$Y(\mathbf{x}) = \ln K_s(\mathbf{x}) = \sum_{k=1}^{K} Y_k(\mathbf{x}) \psi_k(\mathbf{x}), \tag{16}$$

where $Y_k(\mathbf{x}) = \ln K_s(\mathbf{x})$ is a spatial random function defined in subdomain D_k, and $\psi_k(\mathbf{x})$ is a deterministic indicator function given as $\psi_k(\mathbf{x}) = 1$ for $\mathbf{x} \in D_k$ and $\psi_k(\mathbf{x}) = 0$ otherwise. Lu and Zhang [25] developed an algorithm for computing eigenvalues and eigenfunctions for the nonstationary hydraulic conductivity field. The procedure can be summarized as follows:

- Equation (13) is solved for each individual zone D_k to obtain eigenvalues $\{\lambda_n^{(k)}, n = 1, 2, \ldots\}$ and eigenfunctions $\{f_n^{(k)}(\mathbf{x}), n = 1, 2, \ldots\}$;
- Extend the domain of $f_n^{(k)}(\mathbf{x})$ from D_k to the entire domain D by defining $f_n^{(k)}(\mathbf{x}) = 0$ for $\mathbf{x} \notin D_k$;
- Merge K sets of eigenvalues together $\{\lambda_n^{(k)}, k = \overline{1, K}, n = 1, 2, \ldots\}$ and sort them in a non-increasing order (denoting the sorted series as λ_k, $k = 1, 2, \ldots$);
- Arrange the set of merged eigenfunctions $\{f_n^{(k)}, k = \overline{1, K}, n = 1, 2, \ldots\}$ based on the sorted eigenvalues and denote the new set of eigenfunctions as $f_k(\mathbf{x})$, $k = 1, 2, \ldots$.

The KL decomposition of the mean-removed stochastic process $Y'(\mathbf{x})$ for both stationary and non-stationary hydraulic conductivity fields can be written as

$$Y'(\mathbf{x}) = \sum_{n=1}^{\infty} \xi_n \sqrt{\lambda_n} f_n(\mathbf{x}). \tag{17}$$

The hydraulic conductivity field conditioned on some direct measurements is a special kind of nonstationary field. Lu and Zhang [21] developed a methodology to efficiently update eigenvalues and eigenfunctions of the covariance of the hydraulic conductivity field. Since the set of eigenfunctions are complete, the basic idea of this updating scheme is to express the conditional eigenfunctions as linear combinations of the unconditional eigenfunctions and derive equations for the coefficients of these linear combinations. By doing so, the problem of finding the eigenvalues and eigenfunctions of a conditional covariance function $C_Y^{(c)}(\mathbf{x}, \mathbf{y})$ reduces to the problem of finding the eigenvalues and eigenvectors of an $N_Y \times N_Y$ symmetric matrix, where N_Y is the number of conditioning points, which in general is small. The computational cost of finding conditional eigenvalues and eigenfunctions in this way is much less than that of directly solving (13). This updating scheme is a critical component of the KLKF method described in Section 6.

5 KL-based Moment Equations

The Kalman filtering method requires to compute the covariance matrix of the system response, i.e., the **P** matrix in (6)-(11). In the EnKF, this is accomplished by running a series of forward modeling using a number of realizations. In the KLKF method, the covariance of head is derived from forward modeling using the KL-based moment method (KLME). The KLME has been described in detail in [21, 22, 39]. For completeness, a brief description is given as follows.

Since the dependent variable $h(\mathbf{x},t)$ is a function of the input variability $\sigma_Y^2(\mathbf{x})$, one may formally express $h(\mathbf{x},t)$ as an infinite series $h(\mathbf{x},t) = \sum_{m=1}^{\infty} h^{(m)}(\mathbf{x},t)$, where the order of each term in this summation is with respect to σ_Y, the standard deviation of Y. We only keep the first two terms in this series, and the KLKF method described here is based on the first-order approximation of the pressure head. By substituting the expansions of $h(\mathbf{x},t)$ and Y into (1), we obtain the governing equations for zeroth-order pressure head $h^{(0)}(\mathbf{x},t)$

$$\nabla \cdot \left[K_G(\mathbf{x})\nabla h^{(0)}(\mathbf{x},t)\right] + g(\mathbf{x},t) = S_s \frac{\partial h^{(0)}(\mathbf{x},t)}{\partial t}, \tag{18}$$

$$h^{(0)}(\mathbf{x},0) = H_0(\mathbf{x}), \qquad \mathbf{x} \in D, \tag{19}$$

$$h^{(0)}(\mathbf{x},t) = H(\mathbf{x},t), \qquad \mathbf{x} \in \Gamma_D, \tag{20}$$

$$K_G(\mathbf{x})\nabla h^{(0)}(\mathbf{x},t) \cdot \mathbf{n}(\mathbf{x}) = -Q(\mathbf{x},t), \qquad \mathbf{x} \in \Gamma_N, \tag{21}$$

and for the first-order pressure head term $h^{(1)}(\mathbf{x},t)$,

$$\nabla \cdot \left[K_G(\mathbf{x})\nabla h^{(1)}(\mathbf{x},t)\right] + g^{(1)}(\mathbf{x},t) = S_s \frac{\partial h^{(1)}(\mathbf{x},t)}{\partial t}, \tag{22}$$

$$h^{(1)}(\mathbf{x},0) = 0, \quad \mathbf{x} \in D, \tag{23}$$

$$h^{(1)}(\mathbf{x},t) = 0, \quad \mathbf{x} \in \Gamma_D, \tag{24}$$

$$K_G(\mathbf{x})\nabla h^{(1)}(\mathbf{x},t) \cdot \mathbf{n}(\mathbf{x}) = Q(\mathbf{x},t)Y'(\mathbf{x}), \quad \mathbf{x} \in \Gamma_N, \tag{25}$$

where K_G is the geometric mean of the K_s, and

$$g^{(1)}(\mathbf{x},t) = S_s Y'(\mathbf{x}) \frac{\partial h^{(0)}(\mathbf{x},t)}{\partial t} + K_G(\mathbf{x})\nabla Y'(\mathbf{x}) \cdot \nabla h^{(0)}(\mathbf{x},t) - g(\mathbf{x},t)Y'(\mathbf{x}). \tag{26}$$

Equations for higher-order mean head and head covariance can be found in [21, 22, 24, 39].

Equations (18)-(21) are the governing equations and initial and boundary conditions for the zeroth-order conditional mean head. In the conventional moment method, the equations for the first-order (in terms of σ_Y^2) head covariance can be derived from (22)-(25) upon multiplying these equations by

$h^{(1)}(\chi, \tau)$ and taking the ensemble mean. It has been shown that the conventional moment method is computationally expensive, especially for large-scale problems [22]. In fact, to solve the pressure head covariance up to the first order (in σ_Y^2) using the conventional moment-equaiton method, it is required to solve the sets of linear algebraic equations with N unknowns (N being the number of nodes in the numerical grid) for $2N$ times: N times for solving the cross-covariance C_{Yh} and N times for the covariance C_h. Solving the pressure head covariance with higher-order corrections is possible, but the computational effort is very demanding. For instance, solving the pressure head covariance up to the second order in terms of σ_Y^2 requires solving sets of linear algebraic equations with N unknowns for N^2 times.

In the KLME method, instead of dealing with $h^{(1)}$ directly from (22)-(26), we further assume that the first-order head can be exprssed as a polynomial expansion in terms of the orthogonal random variables ξ_m:

$$h^{(1)}(\mathbf{x}, t) = \sum_{m=1}^{\infty} \xi_m h_m^{(1)}(\mathbf{x}, t), \tag{27}$$

where $h_m^{(1)}$ are deterministic, first-order head with mode m. After substituting this expansion and the KL decomposition of the log hydraulic conductivity, i.e., (17), into (22)-(26) and recalling the orthogonality of random variables ξ_m, we obtain sets of governing equations for deterministic coefficients $h_n^{(1)}$:

$$\nabla \cdot \left[K_G(\mathbf{x}) \nabla h_n^{(1)}(\mathbf{x}, t)\right] + g_n^{(1)}(\mathbf{x}, t) = S_s \frac{\partial h_n^{(1)}(\mathbf{x}, t)}{\partial t}, \tag{28}$$

$$h_n^{(1)}(\mathbf{x}, 0) = 0, \quad \mathbf{x} \in D, \tag{29}$$

$$h_n^{(1)}(\mathbf{x}, t) = 0, \quad \mathbf{x} \in \Gamma_D, \tag{30}$$

$$K_G(\mathbf{x}) \nabla h_n^{(1)}(\mathbf{x}, t) \cdot \mathbf{n}(\mathbf{x}) = Q(\mathbf{x}, t) \sqrt{\lambda_n} f_n(\mathbf{x}), \quad \mathbf{x} \in \Gamma_N, \tag{31}$$

where

$$g_n^{(1)}(\mathbf{x}, t) = \left[S_s \frac{\partial h^{(0)}(\mathbf{x}, t)}{\partial t} - g(\mathbf{x}, t)\right] \sqrt{\lambda_n} f_n(\mathbf{x}) + K_G(\mathbf{x}) \sqrt{\lambda_n} \nabla f_n(\mathbf{x}) \cdot \nabla h^{(0)}(\mathbf{x}, t). \tag{32}$$

Here the source term $g_n^{(1)}$ and the flux bounday term in the right side of (31) represent the effect of heterogeneity of the hydraulic conductivity. Note that (28)-(32) are driving by terms that are proportional to $\sqrt{\lambda_n}$. Since $\sqrt{\lambda_n}$ decreases quickly, the first-order head $h^{(1)}$ in (27) can be approximated by a limited number of terms, say M terms. It is worthy to point out that the zeroth-order head equation and the M sets of equations for coefficients $h_n^{(1)}$ (totally there are $M+1$ sets of equations) have the exactly same structure as the original flow equation. By changing the input parameters, the KL-based moment equations can be solved easily with existing flow simulators, such as MODFLOW [15].

Once the coefficients $h_n^{(1)}$, $n = \overline{1, M}$, are solved, we only need to store an $M \times N$ matrix rather than an $N \times N$ full matrix C_h. The head covariance as well as the cross-covariance between the head and the log hydraulic conductivity, which are required in the data assimilation process, can be approximated up to first order in terms of σ_Y^2 whenever needed:

$$C_{Yh}(\mathbf{x}; \mathbf{y}, \tau) = \sum_{m=1}^{M} \sqrt{\lambda_m} f_m(\mathbf{x}) h_m^{(1)}(\mathbf{y}, \tau), \tag{33}$$

$$C_h(\mathbf{x}, t; \mathbf{y}, \tau) = \sum_{m=1}^{M} h_m^{(1)}(\mathbf{x}, t) h_m^{(1)}(\mathbf{y}, \tau). \tag{34}$$

The accuracy of the KLKF method highly depends on how well one can approximate C_{Yh} and C_h from (33)-(34). Zhang and Lu [2004] showed that, when the unconditional variability σ_Y^2 is small, these first-order approximations are close to the true solutions. However, if the variability σ_Y^2 is large (say, $\sigma_Y^2 > 2.0$), higher-order corrections are needed unless the correlation length is small. In the latter case, first-order solutions are sufficiently accurate, although in this case, the number of modes should be relatively large [*Zhang and Lu*, 2004]. In general, the number of required modes depends on the dimensionless size (in terms of the correlation length) of the domain. It is our experience that 100 modes are enough for most of the cases we examined.

6 KL-based Data Assimilation Methodology

The data assimilation process consists of a series of updating steps, each of which represents the time when observations become available and/or the updating process is operated. In this section, the time symbol t is suppressed, because the discussion is based on any fixed assimilation step. Similar to the EnKF, the KLKF requires to compute covariance functions C_Y, C_{Yh}, and C_h, as presented in (12), (33), and (34). This implies that the eigenvalues and eigenfunctions, as well as the first-order term $h_n^{(1)}$ have to be updated at each assimilation step, which in turn requires updating the mean hydraulic conductivity field and the zeroth- and first-order head terms. Certainly, because of conditioning, at each updating step, the covariance function of the log hydraulic conductivity is nonstationary and its eigenvalues and eigenfunctions have to be solved numerically. It is well-known that solving eigenfunctions numerically is computationally very expensive for large-scale problems. For this reason, Zhang et al. [40] developed an algorithm that requires solving the eigenvalue problem at the first time step and then efficiently updating eigenvalues and eigenfunctions at the sequential assimilation steps.

6.1 Updating Mean Hydraulic Conductivity Field

At each update step, for the given observations described in Section 2, the updated mean and covariance of Y upon incorporating new observations can be derived from the cokriging technique:

$$Y^{(0)(c)}(\mathbf{x}) = Y^{(0)}(\mathbf{x}) + \sum_{i=1}^{N_Y} \alpha_i(\mathbf{x}) \left[Y_{obs}(\mathbf{x}_i) - Y^{(0)}(\mathbf{x}_i) \right] + \sum_{i=1}^{N_h} \beta_i(\mathbf{x}) \left[h_{obs}(\chi_i) - h^{(0)}(\chi_i) \right], \tag{35}$$

and

$$C_Y^{(c)}(\mathbf{x}_i, \mathbf{y}_j) = C_Y(\mathbf{x}_i, \mathbf{y}_j) - \sum_{n=1}^{N_Y} \alpha_n(\mathbf{x}_i) C_Y(\mathbf{x}_n, \mathbf{y}_j) - \sum_{n=1}^{N_h} \beta_n(\mathbf{x}_i) C_{Yh}(\mathbf{y}_j, \chi_n), \tag{36}$$

where the quantities with (without) superscript (c) stands for the values after (before) incorporating observations at this time step, and $\alpha_i(\mathbf{x})$ and $\beta_i(\mathbf{x})$ are weighting functions, representing the relative importance of each measurement $Y_{obs}(\mathbf{x}_i)$ and $h_{obs}(\chi_i)$ in predicting the value of $Y^{(0)(c)}(\mathbf{x})$ at location $\mathbf{x}$. The observations $Y_{obs}(\mathbf{x}_i)$ and $h_{obs}(\chi_i)$ may include noises:

$$Y_{obs}(\mathbf{x}_i) = Y^t(\mathbf{x}_i) + \zeta_i \epsilon_Y, \quad i = \overline{1, N_Y} \tag{37}$$

$$h_{obs}(\chi_i) = h^t(\chi_i) + \zeta_i \epsilon_h, \quad i = \overline{1, N_h} \tag{38}$$

where $Y^t(\mathbf{x}_i)$ and $h^t(\chi_i)$ are unknown true values at observation locations, ζ_i are Gaussian random variables with zero mean and unit variance, and ϵ_Y and ϵ_h are the standard deviations of measurements errors of the log hydraulic conductivity $Y(\mathbf{x})$ and pressure head $h(\chi)$, which are assumed to be known.

The weighting functions in (35) are solutions of the following cokriging equations:

$$\sum_{i=1}^{N_Y} \alpha_i(\mathbf{x}) C_Y(\mathbf{x}_i, \mathbf{x}_j) + \sum_{i=1}^{N_h} \beta_i(\mathbf{x}) C_{Yh}(\mathbf{x}_j, \chi_i) = C_Y(\mathbf{x}, \mathbf{x}_j), \ j = \overline{1, N_Y}, \tag{39}$$

$$\sum_{i=1}^{N_Y} \alpha_i(\mathbf{x}) C_{Yh}(\mathbf{x}_i, \chi_j) + \sum_{i=1}^{N_h} \beta_i(\mathbf{x}) C_h(\chi_i, \chi_j) = C_{Yh}(\mathbf{x}, \chi_j), \ j = \overline{1, N_h}. \tag{40}$$

Note that coefficients α_i and β_i are location-dependent, which means that equations (39)-(40) need to be solved for N times, where N is the number of nodes in the domain. Follow [21], α_i and β_i can be directly related to the eigenvalues, eigenfunctions, and the first-order head $h^{(1)}$. Since the set of eigenfunctions (no matter they are unconditional or conditional) is complete, $\alpha_i(\mathbf{x})$ and $\beta_i(\mathbf{x})$ can be expanded on the basis of these eigenfunctions:

$$\alpha_i(\mathbf{x}) = \sum_{k=1}^{\infty} \alpha_{ik} f_k(\mathbf{x}), \tag{41}$$

$$\beta_i(\mathbf{x}) = \sum_{k=1}^{\infty} \beta_{ik} f_k(\mathbf{x}). \tag{42}$$

Substituting (41)-(42) into (39)-(40), multiplying $f_m(\mathbf{x})$ on the both sides of the resulted equations, and integrating the equations with respect to $\mathbf{x}$ over the domain D, we obtain equations for α_{im} and β_{im}

$$\sum_{i=1}^{N_Y} \alpha_{im} C_Y(\mathbf{x}_i, \mathbf{x}_j) + \sum_{i=1}^{N_h} \beta_{im} C_{Yh}(\mathbf{x}_j, \chi_i) = \lambda_m f_m(\mathbf{x}_j), \quad j = \overline{1, N_Y}, \tag{43}$$

$$\sum_{i=1}^{N_Y} \alpha_{im} C_{Yh}(\mathbf{x}_i, \chi_j) + \sum_{i=1}^{N_h} \beta_{im} C_h(\chi_i, \chi_j) = \sqrt{\lambda_m} h_m^{(1)}(\mathbf{x}_j), \quad j = \overline{1, N_h}, \tag{44}$$

where the cross-covariance C_{Yh}, auto-covariance of pressure head C_h, and auto-covariance of the log hydraulic conductivity C_Y are given by (33), (34), and (12), respectively. All these auto- and cross-covariance functions depend on simulated time and thus need to be updated at each assimilation step. It should be noted that (43)-(44) are solved only M times (in general $M << N$), where M is the number of modes needed to approximate Y with a desired accuracy. In addition, solving α_{im} and β_{im} will facilitate updating eigenvalues and eigenfunctions, as described later.

6.2 Updating Pressure Head Fields

The zeroth-order pressure head can be updated with both types of observations in the same manner:

$$\begin{aligned} h^{(0)(c)}(\mathbf{x}) = h^{(0)}(\mathbf{x}) &+ \sum_{i=1}^{N_Y} \mu_i(\mathbf{x}) \left[Y_{obs}(\mathbf{x}_i) - Y^{(0)}(\mathbf{x}_i) \right] \\ &+ \sum_{i=1}^{N_h} \eta_i(\mathbf{x}) \left[h_{obs}(\chi_i) - h^{(0)}(\chi_i) \right], \end{aligned} \tag{45}$$

where $\mu_i(\mathbf{x})$ and $\eta_i(\mathbf{x})$ are subject to:

$$\sum_{i=1}^{N_Y} \mu_i(\mathbf{x}) C_Y(\mathbf{x}_i, \mathbf{x}_j) + \sum_{i=1}^{N_h} \eta_i(\mathbf{x}) C_{Yh}(\mathbf{x}_j, \chi_i) = C_{Yh}(\mathbf{x}_j, \mathbf{x}), \ j = \overline{1, N_Y}, \tag{46}$$

$$\sum_{i=1}^{N_Y} \mu_i(\mathbf{x}) C_{Yh}(\mathbf{x}_i, \chi_j) + \sum_{i=1}^{N_h} \eta_i(\mathbf{x}) C_h(\chi_i, \chi_j) = C_h(\mathbf{x}, \chi_j), \ j = \overline{1, N_h}. \tag{47}$$

Coefficients α_{im}, β_{im}, μ_i, and η_i in (43)-(44) and (46)-(47) can be computed through solving a set of linear algebraic equations with the same coefficient matrix. However, the condition number of this coefficient matrix could be extremely large, especially after several assimilation steps, because assimilating observations will result in a great reduction of (co)variance. Due to the ill-conditioned matrix, the truncation error will easily be amplified reflecting on the anomalies. Dietrich and Newsam [5] analyzed the cause of the ill-conditioning and proposed that adding a relaxation term (or an explicit error matrix) to the coefficient matrix can resolve this problem. In general, a relatively large relaxation term slows down the rate of convergence but with the price of losing information, while a small value may lead to numerical instability. The error matrix can be obtained through a maximum likelihood approach for the purpose of improving the conditioning and minimizing the loss of information. Yeh et al. [37] added a relaxation term to the diagonal components of the matrix to reduce the condition number of the matrix, and the relaxation term is a fraction of the maximum value of the coefficient matrix. In the example shown in Section 7, a constant relaxation term is added to the diagonal components of C_h, which will be discussed further with illustrative examples.

Similarly to the zeroth-order head term, the first-order pressure head is updated by:

$$h^{(1)(c)}(\mathbf{x}) = h^{(1)}(\mathbf{x}) - \sum_{i=1}^{N_Y} \mu_i(\mathbf{x}) Y'(\mathbf{x}_i) - \sum_{i=1}^{N_h} \eta_i(\mathbf{x}) h^{(1)}(\boldsymbol{\chi}_i). \tag{48}$$

Because both Y' and $h^{(1)}$ can be expanded based on ξ_m, by substituting (17) and (27) into (48), instead of updating $h^{(1)}$ directly, the coefficients $h_m^{(1)}$ can be updated as:

$$h_m^{(1)(c)}(\mathbf{x}) = h_m^{(1)}(\mathbf{x}) - \sum_{i=1}^{N_Y} \mu_i(\mathbf{x}) \sqrt{\lambda_m} f_m(\mathbf{x}_i) - \sum_{i=1}^{N_h} \eta_i(\mathbf{x}) h_m^{(1)}(\boldsymbol{\chi}_i). \tag{49}$$

It can be shown that to the first order, (49) recovers the usual cokriging equation for the head covariance similar to (36).

6.3 Updating Eigenvalues and Eigenfunctions

Because of the non-stationality of the covariance matrix given in (36), the conditional eigenvalues and eigenfunctions have to be solved numerically. Here we follow the method in [21] with some modification to incorporate the influence of the pressure head measurements. By definition, the eigenvalues $\lambda_m^{(c)}$ and their corresponding eigenfunctions $f_m^{(c)}$ can be solved from (13) upon replacing C_Y by the conditional covariance function $C_Y^{(c)}$. Because of high computational cost in solving (13), this equation will be solved only at the first assimilation step and the eigenvalues and eigenfunctions for any sequential step will be derived from the results at the previous step. Since the set of eigenfunctions $f_m(\mathbf{x})$ computed at the previous assimilation step is complete, the eigenfunctions $f_m^{(c)}(\mathbf{x})$ at the current step (after assimilating new measurements) can be expanded in terms of $f_m(\mathbf{x})$ as:

$$f_m^{(c)}(\mathbf{x}) = \sum_{p=1}^{M} d_{mp} f_p(\mathbf{x}), \quad m = \overline{1, M} \tag{50}$$

where the coefficient matrix $\mathbf{D} = (d_{mp})_{M\times M}$ is to be determined. Substituting this expression and (36) into (13), multiplying $f_m(\mathbf{y})$ on the both sides of the derived equation, and integrating it with respect to $\mathbf{y}$ over the domain D, yields:

$$\lambda_m d_m - \sum_{k=1}^{M} \left(\sum_{i=1}^{N_Y} \alpha_{ik} \lambda_m f_m(\mathbf{x}_i) + \sum_{i=1}^{N_h} \beta_{ik} h_m^{(1)}(\chi_i) \right) d_k = \lambda_m^{(c)} d_m, m = \overline{1, M} \tag{51}$$

It can also be expressed in a succinct matrix form as:

$$\left(\mathbf{A} - \lambda^{(c)}\mathbf{I}\right)\mathbf{D} = 0, \tag{52}$$

where

$$a_{km} = \lambda_m \delta_{km} - \sum_{k=1}^{M} \left(\sum_{i=1}^{N_Y} \alpha_{ik} \lambda_m f_m(\mathbf{x}_i) + \sum_{i=1}^{N_h} \beta_{ik} h_m^{(1)}(\chi_i) \right) \tag{53}$$

are components of $\mathbf{A} = (a_{km})_{M\times M}$, $\lambda^{(c)} = diag(\lambda_1^{(c)}, \ldots, \lambda_M^{(c)})$, and $\mathbf{I}$ is an $M\times M$ identical matrix. Therefore, the problem of finding the eigenvalues and eigenfunctions of a nonstationary covariance matrix $C_Y^{(c)}(\mathbf{x}, \mathbf{y})$ of size $N \times N$ reduces to the problem of finding the eigenvalues $\lambda^{(c)}$ and eigenvectors d of an $M \times M$ matrix for M times, where N is the number of grid nodes, and M is the number of modes. Note that the number of grid nodes N is usually much larger than the number of modes M. Once $\mathbf{D}$ is solved, conditional eigenvector $f_m^{(c)}$ corresponding to each conditional eigenvalue $\lambda_m^{(c)}$ can be constructed using (50). The updated eigenvalues and eigenvectors as well as the updated pressure head terms in (45) and (49) are fed into the KLME forward model. Since directly solving the Fredholm equation is computationally expensive, the proposed algorithm for updating conditional eigenvalues and eigenfunctions at each updating step has a significant advantage, and we only need to solve the Fredholm equation at the first time step rather than at each assimilation step.

6.4 Karhunen-Loève Based Kalman Filter

The methodology described in the previous section has a different format than the traditional Kalman filter, where the Kalman gain and the updating step are formulated as:

$$\mathbf{G} = \mathbf{P}^f\mathbf{H}^T[\mathbf{H}\mathbf{P}^f\mathbf{H}^T + \mathbf{R}]^{-1}, \tag{54}$$

$$\mathbf{S}^a = \mathbf{S}^f + \mathbf{G}[\mathbf{d} - \mathbf{H}\mathbf{S}^f], \tag{55}$$

where $\mathbf{K}$ is the Kalman gain, $\mathbf{H}$ is the observation operator, $\mathbf{R}$ is the observation error covariance matrix, $\mathbf{d}$ is the observation vector, $\mathbf{S}^f$ is the forecast state vector, $\mathbf{S}^a$ is the updated state vector, and $\mathbf{P}^f$ is the covariance matrix of the forecast state vector. In fact, the cokriging based updating scheme is exactly the same as the traditional Kalman filter. If we define the state vector as $\mathbf{S} = (Y(\mathbf{y}_1), \ldots, Y(\mathbf{y}_N), h(\mathbf{y}_1), \ldots, h(\mathbf{y}_N))^T$, the Kalman gain can be constructed with the coefficients $\alpha_i(\mathbf{x})$, $\beta_i(\mathbf{x})$, $\mu_i(\mathbf{x})$, and $\eta_i(\mathbf{x})$ described before:

$$\mathbf{G} = \begin{bmatrix} \mathbf{G}_\alpha & \mathbf{G}_\beta \\ \mathbf{G}_\mu & \mathbf{G}_\eta \end{bmatrix}_{2\times N, N_Y+N_h} \tag{56}$$

where $\mathbf{G}_\alpha = (\alpha_1, \ldots, \alpha_{N_Y})$, $\mathbf{G}_\beta = (\beta_1, \ldots, \beta_{N_h})$, $\mathbf{G}_\mu = (\mu_1(, \ldots, \mu_{N_Y})$, and $\mathbf{G}_\eta = (\eta_1, \ldots, \eta_{N_h})$. Note that each of α_i, β_i, μ_i, and η_i is a vector of $N \times 1$.

The major difference between the KLKF approach and the cokriging method is that the KLKF is a sequential or inline method, while cokriging is a statistical interpolation method, which only operates at a fixed time. Sometimes cokriging is performed iteratively to account for the possible nonlinear effects in order to obtain a reasonable estimation, but the iteration is still based on a certain time [37]. The KLKF uses the KLME method for advancing the system with time, and incorporates the observations at the time when they become available and update the system at the same time. After the current assimilation step, the updated system responses are taken as the initial conditions for the next forward step. The updated model that runs until the next set of observations become available, at which the updating step will be performed again.

7 Illustrative Examples

In this section, a synthetic two-dimensional example is used to demonstrate the applicability of the KLKF method in estimating the hydraulic conductivity field by assimilating both pressure head and hydraulic conductivity measurements. The results are then compared with those from the EnKF method in terms of both the computational cost and accuracy [40].

The flow domain is a square of size $L_x = L_y = 800$ [L] (where L is any consistent length unit), uniformly discretized into 40×40 square elements, as shown in Figure 1. A pumping well and an injection well are placed at (240 [L], 160 [L]) and (540 [L], 560 [L]), respectively, with a volumetric flow rate of 150 [L^3/day]. Both wells are active throughout the entire simulation period with constant flow rates. The two lateral boundaries are no-flow boundaries, while the left and the right are Dirichlet boundaries with prescribed pressure head of 202 [L] and 198 [L], respectively. Storage coefficient is assumed to be a

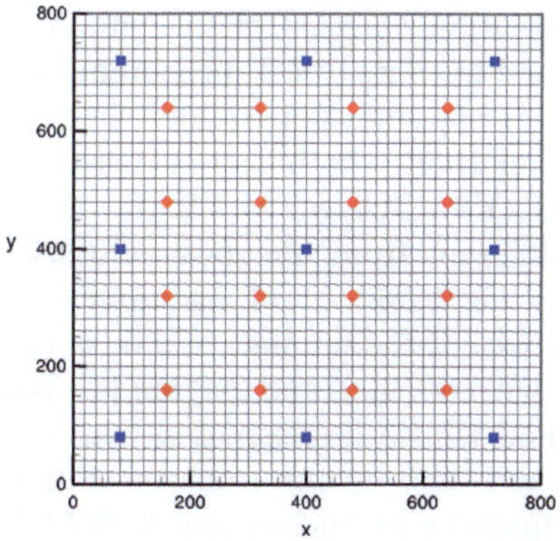

Fig. 1. The flow domain and the observation locations for $\ln K_s$ (9 squares) and h (both squares and diamonds symbols).

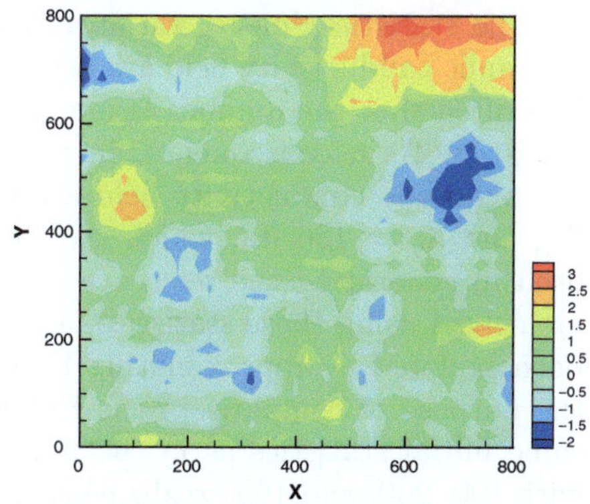

Fig. 2. The reference log hydraulic conductivity field.

constant and taken as 0.0001. The log hydraulic conductivity field is treated as a spatially-correlated Gaussian random field with zero mean and unit variance. The unconditional hydraulic conductivity field is also assumed to be second-order stationary characterized by a separable exponential covariance function

$$C_Y(\mathbf{x}_1, \mathbf{x}_2) = C_Y(x_1, y_1; x_2, y_2) = \sigma_Y^2 \exp\left[\frac{|x_1 - x_2|}{\lambda_x} - \frac{|y_1 - y_2|}{\lambda_y}\right], \quad (57)$$

where $\sigma_Y^2 = 1.0$ is the unconditional variance of Y, and $\lambda_x = 200$ [L] and $\lambda_y = 100$ [L] are correlation lengths in x and y directions, respectively.

In this synthetic example, an unconditional realization of the log hydraulic conductivity field is generated under given statistics using a random field generator based on the KL decomposition [39]. This field is then considered as the "true" field, called the reference field, as shown in Figure 2. Nine samples are taken from this reference field at selected locations as shown in Figure 1 (in blue) and these samples are considered as direct measurements of the log hydraulic conductivity field. A forward transient simulation is conducted using the reference hydraulic conductivity field. For this model setup, the fluid flow reaches steady state at about $t = 10$ [day]. This period is chosen as the duration of the total simulation time, which is then subdivided into 50 equally-sized time intervals with a size of 0.2 [day]. Twenty-five pressure head measurements are then taken at selected locations (both blue and red points in Fig. 1) at elapsed time $t = 0.2 + 0.6k$, $k = 0, 1, \ldots, 16$. It is assumed that the hydraulic conductivity measurements are error-free, while the pressure head observations are noisy and the measurement error follows a normal distribution $N(0, 2.5 \times 10^{-3})$. The hydraulic conductivity measurements are assimilated with the first set of pressure head measurements at $t = 0.2$ [day], and after that only pressure head measurements are assimilated every 0.6 day.

The statistics (mean, variance, and correlation lengths) of the unknown hydraulic conductivity field are usually inferred from its direct measurements. It is often that the number of measurements is not sufficient to infer the true statistics. As a consequence, in the KLKF method, the initial statistics of the hydraulic conductivity field are taken as $\langle Y \rangle = 0.0$, $\sigma_Y^2 = 1.2$, $\lambda_x = 220$ [L], and $\lambda_y = 120$ [L], which are slightly different from the those statistics used in generating the reference field. Initial eigenvalues and eigenfunctions are solved using these statistics, and the impact of the number of modes has been explored by approximating Y and $h^{(1)}$ with 50, 100, and 200 modes. For the purpose of comparison, the EnKF method is also implemented with the same initial statistics as in the KLKF method, and results from the EnKF with 100, 200, and 1000 realizations are compared with those from the KLKF to assess the accuracy and efficiency of the KLKF method.

7.1 Comparison of Accuracy

Two measures are commonly used to evaluate the accuracy of the Kalman filtering schemes. The root mean square error (RMSE) is used to compare the estimated Y field with the reference field, which is given as:

$$RMSE = \sqrt{\frac{1}{N}\sum_{i=1}^{N}\left[Y^*(\mathbf{x}_i) - Y^t(\mathbf{x}_i)\right]^2}, \tag{58}$$

where Y^* stands for the estimated value, Y^t stands for the true reference value, and N is the number of grid nodes.

Another measure is ensemble spread, which is the averaged variation of the ensemble and is defined as

$$Spread = \sqrt{\frac{1}{N}\sum_{i=1}^{N} VAR(\mathbf{x}_i)}, \tag{59}$$

where $VAR(\mathbf{x}_i)$ is the ensemble variance. If the EnKF estimates the uncertainty of the state vector properly, the ensemble spread should be close to the RMSE.

Figure 3(a) compares the RMSE of the KLKF and the EnKF methods with a various number of realizations (or modes). The KLKF is operated with a relaxation term $\epsilon = 0.3$ for the cases with 50, 100, and 200 modes. It is seen from the figure that with 100 principal modes the KLKF can properly propagate the statistics of the real randomness, and the estimate is as good as that from the EnKF method with 1000 realizations. The result from only 50 modes is not very stable, but it is comparable with the EnKF with 200 realizations. The KLKF with 100 and 200 modes give similar behavior of the RMSE, which means that adding more modes beyond 100 modes may not

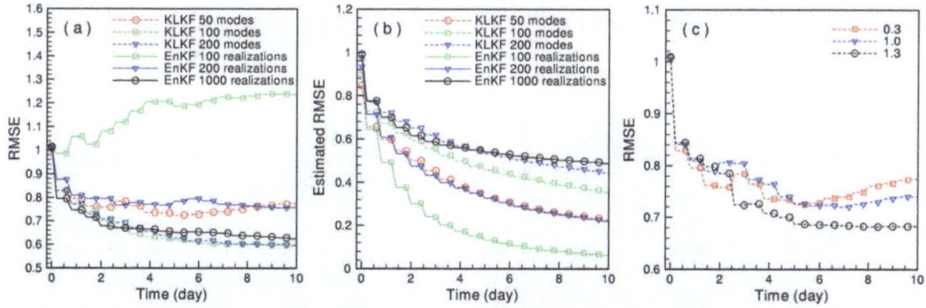

Fig. 3. (a) RMSE and (b) Ensemble spread of the KLKF and the EnKF as a function of time for various number of modes and ensemble sizes.

significantly improve the results. It is important to note that for this case the EnKF with 100 realizations begins to diverge after the first assimilation step.

Figure 3(b) illustrates the ensemble spread of the EnKF with respect to time, in comparison to the uncertainty estimated by the KLKF. Comparing the values shown in Figure 3(b) with the corresponding values in Figure 3(a), the ensemble spread systematically underestimates the real deviation due to the limited size of the ensemble numbers. For instance, the EnKF with 100 realizations fails to reduce the RMSE after the first assimilation step, while the ensemble spread keeps decreasing, indicating that the ensemble is converging to a wrong solution. The estimated uncertainty of the KLKF with 200 modes is equivalent to that of the EnKF with 1000 realizations, which results in the similarity of their corresponding RMSE shown in Figure 3(a). However, the RMSEs of the KLKF with 200 modes and 100 modes are close, and the KLKF with 200 modes has a better estimate of the real deviation, which shows the potential to further incorporate new observations.

The relaxation term used in the KLKF reduces the condition number of the coefficient matrix in (43)-(44) and (46)-(47), and therefore improves stability of the KLKF solution. Numerical experiments [40] show that the performance of the KLKF is sensitive to the choice of relaxation term when the number of modes is small, for instance 50. However, reasonable results can be achieved by cautiously choosing the relaxation term for the KLKF with 50 modes (shown in Figure 3(c)). With larger number of modes, the KLKF has satisfactory performance as long as the relaxation term is changing within a reasonable range. Sensitivity of model results on the relaxation term has been investigated in [40]. In the KLKF approach, the log hydraulic conductivity measurements are used only at the first assimilation step, and the relaxation term is only added to the diagonal terms of the pressure head covariance matrix. If the direct measurements are assimilated with the pressure head observations at every assimilation step, it is expected that a relaxation term needs to be added to the diagonal components of both log hydraulic conductivity and pressure head covariance matrices.

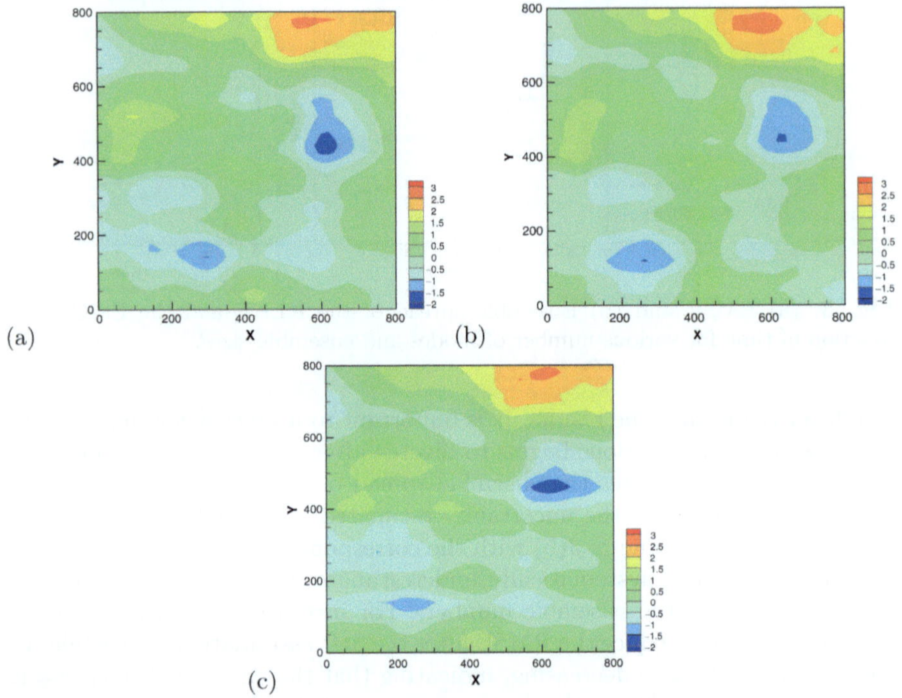

Fig. 4. The estimated mean $\ln K_s$ fields (a) KLKF with 100 modes, (b) KLKF with 200 modes, and (c) EnKF with 1000 realizations.

Zhang et al. [40] also investigated the impact of incorrect initial statistics (comparing to those used in generating the reference field) on the final assimilation results. Their study indicates that the impact is minor if the number of modes included in the KLKF method is enough, for example, 100 modes.

The contours of the log hydraulic conductivity fields estimated from the KLKF method with different numbers of modes are shown in Figure 4. Also compared in the figure is the estimated field using the EnKF method with 1000 realizations. Compared to the reference field (Fig. 2), the contours from the KLKF are smoother. The KLKF with 50 modes (not shown here) fails to catch the right locations of the major pattern of the reference field, while the KLKF with 100 modes almost identifies every primary structure. The contour of the EnKF also recovers the major pattern with more details than the KLKF fields. However, these small features do not exactly replicate those of the reference field. The contours of the associated log hydraulic conductivity variance are shown in Figure 5. Comparing the KLKF results with 100 modes and 200 modes reveals that the variance of the estimated $\ln K_s$ field is higher if more modes are included in data assimilation. The major patterns of estimated fields from the KLKF and the EnKF are close. For the EnKF,

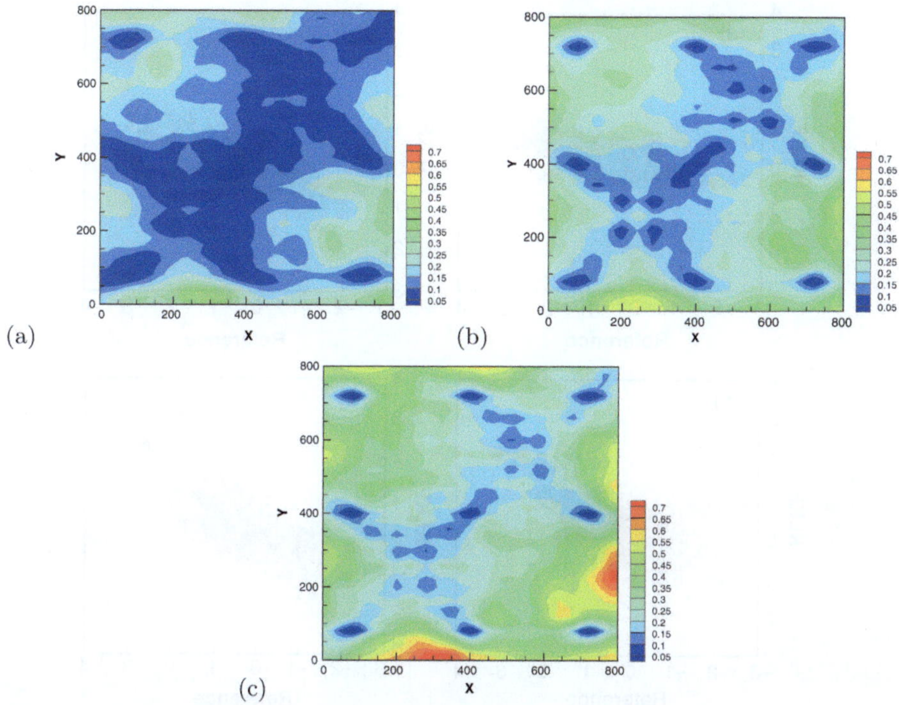

Fig. 5. The estimated $\ln K_s$ variance (a) KLKF with 100 modes, (b) KLKF with 200 modes, and (c) EnKF with 1000 realizations.

although the log hydraulic conductivity measurements are only assimilated at the first assimilation step, the variances remain lowest at those locations. However, this feature is not very obvious in the KLKF, which may be attributed to the effect of the relaxation term.

Figure 6 compares the reference field and the estimated log hydraulic conductivity field from the KLKF with 100 modes at several elapsed times. The figure shows that with time (and available observations), the estimated field becomes closer to the reference field.

In order to test the predictability of the model using the estimated hydraulic conductivity field, we run a deterministic flow simulation from the $t = 0$ [day] to $t = 20$ [day] using the final estimated log hydraulic conductivity field (at 10th day) from the KLKF with 100 modes as the initial input. The calibrated pressure head at $t = 5$ [day] and the predicted head at $t = 20$ [day] are illustrated in Figure 7 (dashed contour lines) as compared to the reference pressure head field (solid contour lines), which is computed using the reference hydraulic conductivity field. Apparently, both head fields match very well.

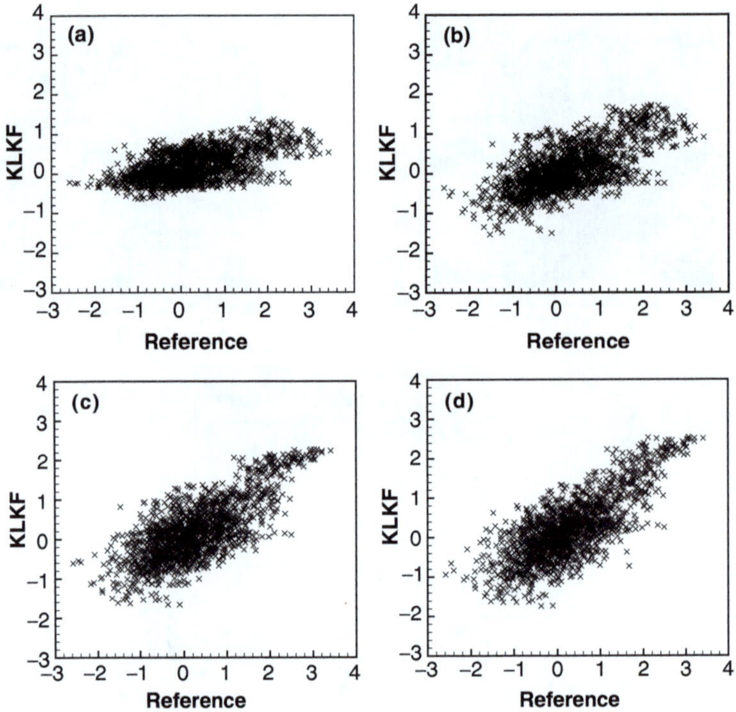

Fig. 6. Comparison between the estimated $\ln K_s$ from the KLKF with 100 modes and the reference at different times. (a) 0.2 day, (b) 2.0 day, (c) 5.0 day and (d) 10.0 day.

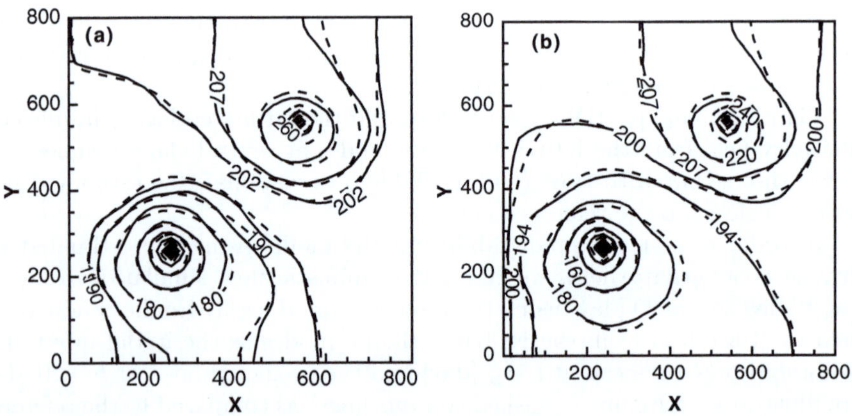

Fig. 7. Comparison of the pressure head fields computed using the reference $\ln K_s$ field (solid lines) and those computed using the estimated mean $\ln K_s$ field from K LKF with 100 modes (dashed lines). (a) pressure head at day 5 and (b) pressure head at day 20.

7.2 Computational Efficiency

If M modes are used in the KLKF, the flow equation needs to be solved for $M+1$ times (once for the zeroth-order head term $h^{(0)}(\mathbf{x})$ and M times for the first-order head terms $h_m^{(1)}(\mathbf{x})$), while for the EnKF with the ensemble size of K, the similar equation needs to be solved for K times. Since the CPU time for solving the pressure head term in the KLKF is about the same as that for one realization in the EnKF method, the computational efficiency for the two methods simply depends on how many modes in the KLKF method (or realizations in the EnKF method) are needed to approximate the statistics of the state. The illustrative example showed that the KLKF can achieve satisfactory estimation with a relatively small computational cost.

8 Summary and Conclusion

The Kalman filter based sequential data assimilation methods have been widely used in solving the inverse problem recently. These methods are capable of updating the system parameters continuously and sequentially with the availability of the measurements of the system responses. In these methods, both the best estimate and the corresponding uncertainty are advanced with time. A major problem associated with these existing Kalman filter based methods is the high computational cost in updating the state error covariance matrix. In this chapter, the Karhunen-Loève based Kalman filter (KLKF) is introduced. The hydraulic conductivity field is treated as a random spatial function and is decomposed using the KL expansion. The pressure head is expanded using the perturbative polynomial expansion. On the basis of these expansions, the higher-order terms are truncated and the KLKF is based on the first-order approximation of the pressure head. The KLKF utilizes only a small number of principal modes to propagate the statistics of the state vector, which greatly reduces the computational cost. The forward step can be solved accurately and efficiently using the Karhunen-Loève based moment-equation method (KLME), which can be solved in parallel using any existing flow model. The data assimilation step is operated based on the state statistics given by the forward step and the observations. A synthetic two-dimensional example shows that the KLKF method is better than the EnKF method in terms of both computational efficiency and accuracy. The example indicates that the estimated conductivity field using the KLKF method with 100 modes is reasonably close to the true reference conductivity field and the pressure head predicted using the estimated field agrees well with the reference head field.

9 Nomenclature

C_h = covariance of pressure head h
C_Y = covariance of $Y = \ln K_s$
C_{Yh} = cross-covariance between Y and h
D = the simulation domain (or its size)
D_m = subdomain
$\mathbf{d}$ = observation vector
$\mathbf{e_1}, \mathbf{e_2}$ = error vectors
$f, f_m, f_n^{(k)}$ = eigenfunction
$\mathbf{G}$ = Kalman gain
g = source/sink
$g_n^{(k)}$ = n^{th} mode of the kth-order source/sink
H = prescribed head at constant head boundary
H_0 = initial head in the domain
$\mathbf{H}$ = observation operator
$h^{(k)}$ = k^{th}-order pressure head
$h_n^{(k)}$ = n^{th} mode of the kth-order pressure head
$\mathbf{I}$ = identical matrix
K = number of subdomains
K_s = hydraulic conductivity
K_G = geometric mean of the hydraulic conductivity
M = number of modes
N = number of grid nodes
N_Y = number of log hydraulic conductivity observations
N_h = number of pressure head observations
$\mathbf{n}$ = an outward unit vector normal to external boundaries
$\mathbf{P}$ = covariance matrix of the state vector
$\mathbf{R_1}, \mathbf{R_2}$ = covariance matrix of error vectors $\mathbf{e_1}$, $\mathbf{e_2}$
$\mathbf{S}$ = state vector
S_s = specific storage
t = time
$\mathbf{x}, \mathbf{y}$ = Cartesian coordinate vectors
$\mathbf{x}_i$ = measurement locations of hydraulic conductivity
Y = log hydraulic conductivity
Y' = zero-mean fluctuation of Y
Y_k = log hydraulic conductivity in zone k
Γ_D = Dirichlet boundary segments
Γ_N = Neumann boundary segments
$\mathbf{\Phi}$ = linear transfer matrix
α_i, β_i = weight coefficients in cokriging estimate of Y
δ_{mn} = Kronecker data
ϵ_Y, ϵ_h = standard deviation of measurments errors on Y and h
ζ = standard Gaussian random variable

$\lambda, \lambda_m, \lambda_n^{(k)}$ = eigenvalue
λ_x, λ_y = correlation lengths in x and y directions
μ_i, η_i = weight coefficients in cokriging estimate of h
ξ_m = orthogonal Gaussian random variables
σ_Y^2 = variance of the log hydraulic conductivity
$\boldsymbol{\chi}_i$ = head measurement locations
ψ_k = indicator function for zone k

Subscript
Obs = observation
en = ensemble

Superscript
(c) = conditioned
f = forecast
u = updated
T = transpose
t = true

References

1. Anderson J L, Anderson S L (1999) Monthly Weather Review 127: 2741-2758
2. Carrera J, Neuman S P (1986) Water Resour. Res. 22(2): 199-210
3. Chen Y, Zhang D (2006) Advances in Water Resour. 29: 1107-1122
4. Dagan G (1985) Water Resour. Res. 21(1): 65-72
5. Dietrich C R, Newsam G N (1989) Stoch. Hydrol. Hydraul. 3: 293-316
6. Evensen G (1994) J. Geophys. Res. 99(C5): 10143-10162
7. Evensen G (2003) Ocean Dynamics 53: 343-367
8. Gelb A (1974) Applied Optimal Estimation, The MIT press
9. Ghanem R, Spanos P D (1991) Stochastic Finite Elements: A Spectral Approach, Springer-Verlag, New York
10. Ghanem R (1998) Water Resour. Res. 34: 2123-2136
11. Ginn T R, Cushman J H (1990) Stoch. Hydrol. Hydraul. 4: 1-26
12. Gomez-Hernanez J J, Sahuquillo A S, Capilla J E (1997) J. Hydrol. 203: 162-174
13. Gu Y, Oliver D S (2004) SPE 89942
14. Gutjahr A, Wilson J (1989) Transport in Porous Media 4: 585-598
15. Harbaugh A W, Banta E R, Hill M C, and McDonald M G (2000) MODFLOW-2000, the U.S. Geological Survey modular ground-water model – User guide to modularization concepts and the Ground-Water Flow Process: U.S. Geological Survey Open-File Report 00-92.
16. Heemink A W, Verlaan M, Segers A J (2001) Monthly Weather Review 129: 1718-1728
17. Houtekamer P L, Mitchell H L (1998) Monthly Weather Review 126: 796-811
18. Hughson D L, Gutjahr A (1998) Stoch. Hydrol. and Hydraul. 12: 155-170
19. Kitanidis P K, Vomvoris E G (1983) Water Resour. Res. 19(3): 677-690
20. Liu N, Oliver D S (2005) J. Petrol. Sci. and Eng. 47: 147-161
21. Lu Z, Zhang D (2004) Adv. in Water Resour. 27(9): 859-874
22. Lu Z, Zhang D (2004) SIAM J. Sci. Comp. 26(2): 558-577
23. Lu Z, Higdon D, Zhang D (2004) Comput. Methods in Water Resour., 1273-1283
24. Lu Z, Zhang D (2006) SPE Journal 11(2), 10.2118/93452-PA, 239-247

25. Lu Z, Zhang D (2007) Multiscale Modeling and Simulation, in press
26. McLaughlin D, Townley L R (1996) Water Resour. Res. 32(5): 1131-1162
27. Mclaughlin D (2002) Adv. in Water Resour. 25: 1275-1286
28. Naevdal G, Johnsen L M, Aanonsen S I, Vefring E H (2003) SPE 84372
29. Neuman S P (1980) Water Resour. Res. 16(2): 331-346
30. Pham D, Verron J, Roubaud (1998) J. Mar. Syst. (16):323-340
31. Sun N-Z, Yeh W W-G (1992) Water Resour. Res. 28(12): 3269-3280
32. Tartakovsky D M, Neuman S P, Lu Z (1999) Water Resour. Res. 35(3): 731-745
33. Verlaan M, Heemink A W (1997) Stoch. Hydrol. Hydraul. 11: 347-368
34. Whitaker J S, Hamill T M (2002) Monthly Weather Review 130: 1913-1924
35. Woodbury A D, Ulrych T J (2000) Water Resour. Res. 36(8): 2081-2093
36. Yeh W W G (1986) Water Resour. Res. 22(2): 95-108
37. Yeh T C, Jin M, Hanna S (1996) Water Resour. Res. 32(1): 85-92
38. Zhang D (2002) Stochastic Methods for Flow in Porous Media: Coping with uncertainties, Academic Press, San Diego, Calif.
39. Zhang D, Lu Z (2003) J. Comput. Physics 194(2): 773–794
40. Zhang D, Lu Z, Chen Y (2007) SPE Journal, 12(1), 10.2118/95277-PA, 108-117.
41. Zimmerman D A, de Marsily G, Gotway C A, Marietta M G, Axness C L, Beauheim R L, Bras R L, Carrera J, Dagan G, Davies P B, Gallegos D P, Galli A, Gomez-Hernandez J, Grindgrod P, Gutjahr A L, Kitanidis P K, LaVenue A M, McLaughlin D, Neuman S P, RamaRao B S, Ravenne C, Rubin Y, (1998) Water Resour. Res. 34(6): 1373-1413

Trajectory-Based Methods for Modeling and Characterization

D.W. Vasco

Berkeley Laboratory, University of California

1 Introduction and Background

Given the increasing data flow associated with characterizing and modeling natural hydrological systems there is a need for efficient data analysis tools. New experimental techniques such as multi-level samplers and crosswell pressure tomography produce extensive sets of hydrologic observations. Furthermore, integrating geophysical and remote sensing data can produce a massive array of measurements. Relating these data to flow properties in the subsurface and using them for characterization can be a computational challenge. In this chapter I discuss trajectory-based methods for integrating transient head, tracer, multi-phase flow, and associated geophysical data. The techniques, which are similar to ray methods in geophysics, are motivated by two methodologies outlined in this chapter. The first methodology, implemented in the frequency-domain, is along the lines of methods associated with elastic and electromagnetic wave propagation [26, 28, 30]. The second approach, the method of multiple scales is somewhat more general, applicable to nonlinear and diffusive propagation, and has been applied to gas dynamics and shock propagation [2]. The techniques described have connections with other methods used in the modeling of fluid flow. For example asymptotic approaches have been used to describe solute transport in the limit of long times, giving for example traveling wave solutions [19, 22, 46, 47]. Furthermore, streamline modeling is a trajectory-based, physically-motivated technique which is extremely useful in modeling tracer transport and multiphase flow [10, 25].

A trajectory-based methodology does offer some computational advantages over purely numerical modeling of flow and transport. Moreover, the methodology also offers insight into the problem of inverse modeling. For example, in the trajectory-based approach the modeling is broken into two distinct steps: (1) the computation of a travel time, (2) the computation of the time-varying amplitude function. This sub-division provides additional flexibility when one turns to inverse modeling. That is, one can use travel times as a datum for characterization. There is some advantage associated with the

D.W. Vasco: *Trajectory-Based Methods for Modeling and Characterization*, Studies in Computational Intelligence (SCI) **79**, 69–103 (2008)
www.springerlink.com

use of travel times. First, the inverse problem associated with travel times is quasi-linear and thus convergence is less of an issue [7]. Furthermore, the inversion of travel times typically involves much less computation than does amplitude inversion. Thus, a useful strategy invokes an initial travel time inversion followed by an amplitude inversion. Trajectory-based modeling also provides, in the form of the trajectories, an easy way to visualize the relationship between observations and the region of the subsurface which influences the observations.

2 Methodology

Natural systems are plagued by a number of characteristics which preclude a straight-forward mathematical analysis. First and foremost, the Earth is highly heterogeneous with properties which may vary by orders of magnitude. Thus, techniques which rely upon constant coefficients or spatially invariant properties are not applicable for modeling processes in the subsurface. Secondly, under commonly encountered conditions, nonlinear behavior is important. This is particularly true for activities of interest such as multiphase flow and finite deformation within the Earth. Thirdly, most natural systems consist of complex, coupled processes which interact over a range of space and time scales. Fourth, many processes are not well characterized and must be described by laboratory-derived, non-analytical methods which are largely empirical. Thus, no closed-form equations are available to describe the specific phenomena.

The preceding factors have motivated many to dispense with analytic solutions in favor of purely numerical approaches [38]. Here I shall pursue semi-analytical solutions because they provide insight into the relationship between observations and properties of the medium. The approach I take relies on a power series representation in terms of a parameter, such as frequency. The series is such that, as the parameter approaches a limiting value, usually zero, higher order terms in the series asymptotically approach zero. Thus, only the first few terms of the series are necessary for an accurate representation. The advantage of an asymptotic approach is that the technique provides solutions even in the presence of three-dimensional, smoothly-varying heterogeneity, and nonlinearity. Furthermore, the components of the solution have useful physical interpretations, which I discuss below.

While asymptotic methods rely on the limiting behavior of an expansion parameter and cannot match the generality of a fully numerical treatment, the methods do have a wide range of applicability. This is not always fully appreciated, and there is the general impression that asymptotic methods are only useful for modeling high-frequency hyperbolic wave propagation [1]. In fact, techniques, such as the method of multiple scales, are general and can be used to model a variety of physical processes in the presence of large-amplitude yet smoothly-varying heterogeneity. I have found that solutions derived using the method of multiple scales represent a wide range of behaviors including diffusive pressure propagation [56], hyperbolic propagation of non-diffusive and non-dispersive tracer [52], as well as mixed-behavior and nonlinear propagation associated with multiphase flow [49]. Recently, I have also used the technique to study broadband electromagnetic wave propagation [51], which can be of a diffusive, hyperbolic or mixed character, depending on the frequency range.

2.1 Frequency Domain Approach

The first approach is based upon an expansion in frequency and follows a Fourier transformation of the governing equation. The Fourier transform is a linear operator given by [4]

$$U(\mathbf{x},\omega) = \frac{1}{2\pi}\int_{-\infty}^{\infty} u(\mathbf{x},t)e^{i\omega t}d\omega. \tag{1}$$

Because of its reliance on the Fourier transform, the technique is applicable to processes governed by linear partial differential equations with spatially varying but time-invariant coefficients. Depending on the application one is generally interested in solutions for either large or small values of frequency, ω. For illustration, I consider the case in which ω is large. An asymptotic solution is a series representation in inverse powers of ω. A fairly general form for an asymptotic solution $u(\mathbf{x},\omega)$ is

$$U(\mathbf{x},\omega) = \exp\left[i\omega r(\mathbf{x}) - \omega^{\alpha} s(\mathbf{x})\right] \sum_{n=0}^{\infty} U_n(\mathbf{x})\omega^{-\lambda_n} \tag{2}$$

where $r(\mathbf{x})$, $s(\mathbf{x})$, and $U_n(\mathbf{x})$ are functions to be determined, α and λ_n are real numbers with $\lambda_{n+1} > \lambda_n$ [15]. For now, disregard the details of the expansion, I shall consider them in the applications below. The main point is that for large ω, the series will be dominated by the first one or two terms. As seen below, using the asymptotic power series one can derive a semi-analytic expression for $U(\mathbf{x},\omega)$ valid in a three-dimensional, heterogeneous medium. The functions $U_n(\mathbf{x})$ are successive amplitude corrections which are required for increasingly higher order accuracy in ω. The functions $r(\mathbf{x})$ and $s(\mathbf{x})$ represent a generalized arrival time or phase, related to the moment at which the disturbance reaches the point $\mathbf{x}$. Note that the series is capable of representing both oscillatory and decaying solutions depending on the presence or absence of $r(\mathbf{x})$ and $s(\mathbf{x})$, respectively. Thus, the method can be applied to both hyperbolic and parabolic differential equations, meaning that it can be used to model non-dispersive tracer transport as well as transient head [24, 28]. One could also maintain such flexibility by setting $r(\mathbf{x})$ equal to zero and allowing $s(\mathbf{x})$ to be complex.

2.2 The Method of Multiple Scales

As described in [2], the method of multiple scales is useful in a variety of settings, from modeling linear hyperbolic wave propagation to the study of nonlinear, dispersive, and diffusive waves. The crucial assumption is that there is a propagating front, separating the material which has not yet experienced the disturbance from the material through which the disturbance has already passed. The front defines a rapid or sharp jump in some quantity or quantities, for example pore pressure, volumetric change, and/or saturation. The definition of 'sharp' is relative to the scale-length of the variation in material properties. I can represent the time and space scale of the front variation by l. Similarly, the variation in background quantities, such as heterogeneity, is over a time and space scale L, where $L \gg l$. I represent the ratio l/L by a dimensionless parameter ϵ and require that $0 < \epsilon \ll 1$. I may define slow variables, in both space and time, in terms of the ratio ϵ:

$$\mathbf{X} = \epsilon^{\alpha}\mathbf{x} \tag{3}$$

$$T = \epsilon^{\alpha}t \tag{4}$$

where α is a rational number. The quantity α is chosen such that the nonlinearity balances the dispersion and dissipation [2]. Formally, an asymptotic solution of a governing equation is a power series representation of the field, in terms of the scale parameter ϵ

$$u(\mathbf{X}, T) = \sum_{n=1}^{\infty} U_n(\mathbf{X}, T, \theta)\epsilon^n. \tag{5}$$

The unknown quantities in equation (5), the phase or travel time $\theta(\mathbf{X}, T)$ and the amplitude corrections $U_n(\mathbf{X}, T, \theta)$, are found by substituting the series into the governing differential equation and examining terms of various orders in ϵ. The low-order components in ϵ are of special interest because they dominate for a relatively sharp front i.e. for $\epsilon = l/L \ll 1$.

3 Application of the Frequency Domain Approach to Transient Head Observations

Using the frequency domain approach I derive a trajectory-based solution for transient head [56]. This solution was used to invert data from two interference tests and image the conductivity distribution associated with a vertical fracture [20, 56].

3.1 The Equation Governing Drawdown

In an inhomogeneous medium the equation describing the space $\mathbf{x}$ and time t evolution of head $h(\mathbf{x}, t)$ is

$$K(\mathbf{x})\nabla^2 h(\mathbf{x},t) + \nabla K(\mathbf{x}) \cdot \nabla h(\mathbf{x},t) = S(\mathbf{x})\frac{\partial h(\mathbf{x},t)}{\partial t} \tag{6}$$

where $K(\mathbf{x})$ denotes the hydraulic conductivity and $S(\mathbf{x})$ denotes the specific storage [3, 13] which has units of inverse length. In the frequency domain, after applying the Fourier transform to equation (6), one has

$$\nabla^2 H(\mathbf{x},\omega) + \mathbf{\Lambda}(\mathbf{x}) \cdot \nabla H(\mathbf{x},\omega) - i\omega\kappa(\mathbf{x})H(\mathbf{x},\omega) = 0 \tag{7}$$

where $\mathbf{\Lambda}$ is the gradient of the logarithm of conductivity,

$$\mathbf{\Lambda}(\mathbf{x}) = \nabla \ln K(\mathbf{x}), \tag{8}$$

which vanishes for constant $K(\mathbf{x})$, and

$$\kappa(\mathbf{x}) = \frac{S(\mathbf{x})}{K(\mathbf{x})} \tag{9}$$

is the inverse of the diffusivity. For large values of ω the head variation will be dominated by κ. Conversely, for small values of ω (low-frequency) the head variation is sensitive to the gradient of the logarithm of conductivity and insensitive to the storage.

3.2 Asymptotic Solutions for Head

As noted above, an asymptotic solution to equation (7) is a power series in inverse powers of ω, a specific case of equation (2),

$$H(\mathbf{x},\omega) = e^{-\sqrt{-i\omega}\sigma(\mathbf{x})}\sum_{n=0}^{\infty}\frac{A_n(\mathbf{x})}{(\sqrt{-i\omega})^n} \tag{10}$$

for $r(\mathbf{x}) = 0, \alpha = 1/2, \lambda_n = n/2, U_n(\mathbf{x}) = A_n(\mathbf{x})\sqrt{-i}^{\,-n}$, and $s(\mathbf{x}) = \sqrt{-i}\sigma(\mathbf{x})$. This form may be deduced on physical grounds, by considering a large argument expansion of the solution to the diffusion equation for a homogeneous medium, a modified Bessel function of zeroth order [58].

In order to obtain explicit expressions for $\sigma(\mathbf{x})$ and $A_n(\mathbf{x})$, the sum (10) is substituted into equation (7). The operators in equation (7) may be applied term by term to the series. The subsitution of the expansion (10) into equation (7) produces an expression containing an infinite number of terms. Each term will contain $\sqrt{-i\omega}$ to some power and I may consider the sets of terms for any given order. However, when ω is large, the first term in the series

$$H(\mathbf{x},\omega) = A_0(\mathbf{x})e^{-\sqrt{-i\omega}\sigma(\mathbf{x})} \tag{11}$$

or its time-domain equivalent, obtained by inverse Fourier transforming equation (11),

$$h(\mathbf{x},t) = A_0(\mathbf{x})\frac{\sigma(\mathbf{x})}{2\sqrt{\pi t^3}}e^{-\sigma^2(\mathbf{x})/4t}, \tag{12}$$

[56, 58], will accurately represent the solution to equation (7). I find that the functions $\sigma(\mathbf{x})$ and $A_0(\mathbf{x})$ are determined by the terms of order $(\sqrt{-i\omega})^2$ and $\sqrt{-i\omega}$ respectively. For this reason, I consider terms of each order in the sub-sections that follow.

3.3 The Eikonal Equation, Trajectories, and the Arrival Time of the Maximum Drawdown

If I consider terms of highest order in $\sqrt{-i\omega}$, those of order $(\sqrt{-i\omega})^2$, I arrive at a scalar, nonlinear, partial differential equation for the phase $\sigma(\mathbf{x})$

$$\nabla\sigma(\mathbf{x}) \cdot \nabla\sigma(\mathbf{x}) - \kappa(\mathbf{x}) = 0. \tag{13}$$

Equation (13), known as the eikonal equation, governs many types of propagation processes [26, 28] and there are efficient numerical methods for its solutions [41]. The eikonal equation relates the phase function $\sigma(\mathbf{x})$ to the flow properties $S(\mathbf{x})$ and $K(\mathbf{x})$, as contained in $\kappa(\mathbf{x})$ [equation (9)].

I may solve equation (13) directly, using the method of characteristics [8]. In the method of characteristics, solutions are developed along particular trajectories, the characteristic curves, which are denoted by $\mathbf{X}(l)$, where l is a parameter signifying position along the curve. The equations for the characteristic curves are a set of four ordinary differential equations

$$\frac{d\mathbf{X}}{dl} = \mathbf{p} \tag{14}$$

$$\frac{d\sigma}{dl} = \sqrt{\kappa} \tag{15}$$

where $\mathbf{p} = \nabla\sigma$ [8]. For a coordinate system with one axis oriented along $\mathbf{p}$ I may write (14) as

$$\frac{dr}{dl} = -p \tag{16}$$

where $p = |\mathbf{p}|$ and r denotes the distance along the axis aligned with $\mathbf{p}$. From equation (15), I can write the phase function as an integral

$$\sigma(\mathbf{x}) = -\int_{\Sigma(\mathbf{x})} \sqrt{\kappa}\, dl \tag{17}$$

where $\Sigma(\mathbf{x})$ is the trajectory from the injection well to the observation point $\mathbf{x}$. The trajectory is found by solving the ray-equations (14) using a numerical technique. The method for determining the trajectory $\mathbf{X}(l)$ from a source point to a given observation point is known as the shooting method for a two-point boundary value problem [37].

A physical interpretation of $\sigma(\mathbf{x})$ is possible, starting from the asymptotic solution $h(\mathbf{x}, t)$, given by equation (12). Differentiating equation (12) with respect to time, and setting the resulting expression to zero, gives a condition for the vanishing of the slope. This condition is satisfied at the peak of the pulse and occurs at the time T_{peak} which is given by $T_{peak} = \sigma^2/6$ [56, 58]. Thus, $\sigma = \sqrt{6T_{peak}}$, the function $\sigma(\mathbf{x})$ is related to the arrival time of the peak pressure associated with a propagating pressure pulse. It has been noted [56] that, for a step function source, $\sigma(\mathbf{x})$ corresponds to the peak of the derivative of the head variation. One can use this relationship to compute the phase function $\sigma(\mathbf{x})$ from the output of a numerical flow simulator. Specifically, by computing the arrival time of the peak value for each grid block one can obtain the distribution of $\sigma(\mathbf{x})$ over the simulation grid. In effect, one is solving the Eikonal equation (13) using the output of a numerical simulation. From the distribution of $\sigma(\mathbf{x})$ it is straight-forward to find the trajectories [see equation (14)] using a second-order Runge-Kutta technique known as Heun's method which is accurate yet simple to implement [55]. Such a hybrid approach is useful when combining trajectory-based inversion techniques with numerical simulation.

3.4 The Transport Equation for the Amplitude Variation of Drawdown

Considering terms of order $\sqrt{-i\omega}$ in the asymptotic expansion (10) I find, after making use of equation (13), the following relationship between $\sigma(\mathbf{x})$ and $A_0(\mathbf{x})$

$$\nabla \cdot \nabla\sigma(\mathbf{x})A_0(\mathbf{x}) + 2\nabla\sigma(\mathbf{x}) \cdot \nabla A_0(\mathbf{x}) + \mathbf{\Lambda}(\mathbf{x}) \cdot \nabla\sigma(\mathbf{x})A_0(\mathbf{x}) = 0. \tag{18}$$

In order to integrate equation (18) I introduce the variable γ such that

$$d\gamma = \frac{ds}{\sqrt{\kappa(\mathbf{x})}} \tag{19}$$

and the unit vector $\mathbf{l}$ which points along $\nabla\sigma$. Equation (18) may be rewritten as an ordinary differential equation along the trajectory Σ

$$2\frac{d\ln(A_0)}{d\gamma} + \frac{d\ln(K)}{d\gamma} + \nabla \cdot \nabla\sigma = 0. \tag{20}$$

I may integrate equation (20) along the trajectory from γ_0 to γ

$$A_0(\gamma) = A_0(\gamma_0)\sqrt{\frac{K(\gamma_0)}{K(\gamma)}}\exp\left(-\frac{1}{2}\int_{\gamma_0}^{\gamma} \nabla \cdot \nabla\sigma d\gamma\right) \tag{21}$$

where $A_0(\gamma_0)$ is the initial pressure amplitude at the source and $K(\gamma_0)$ is the conductivity at the source.

3.5 Trajectory-Based Characterization and Transient Pressure Data

Up to this point I have addressed the forward problem in which observable quantities, such as hydraulic head, are computed given a model of sub-surface flow properties. Now I wish to discuss the inverse problem in which a model of sub-surface flow properties is obtained from a set of observations. Typically, due to factors such as non-uniqueness, non-linearity, and errors in the observations, this is a much more difficult problem [35, 44]. For example, due to the nonlinearity of the inverse problem, repeated solution of the forward problem is required. That is, one must either find a model which fits the observations using a stochastic method such as simulated annealing [12] or use an iterative optimization method such as the Levenburg-Marquardt algorithm. Iterative optimization methods generally converge much more rapidly than stochastic techniques. However, iterative updating schemes require the calculation of model parameter sensitivities or the Jacobian matrix [5, 42, 60]. Model parameter sensitivities relate small changes in subsurface properties to small changes in the time-lapse measurements. More formally, if the conductivity in the jth grid block of a reservoir model is perturbed by δK_j and the corresponding perturbation in the ith observation is δd_i, the sensitivities (M_{ij}) relate the perturbations

$$\delta d_i = \sum_j M_{ij} \delta K_j. \tag{22}$$

Calculating sensitivities is often a tremendous computational undertaking and a bottleneck for most inversion algorithms. Techniques for computing sensitivities range from purely numerical perturbation methods to adjoint-state and the sensitivity equation techniques. These techniques involve either considerable computation or extensive programming. The computational burden is often compounded by the tremendous amount of data that new measurement techniques and geophysical surveys can provide. Trajectory-based methods provide semi-analytic expressions for travel times and amplitudes. These expressions may be used to derive semi-analytic sensitivity estimates [52].

Arrival Time Matching

In order to conduct an iterative inversion I need to relate a perturbation in flow properties to a perturbation in the observations, in this case a perturbation in the arrival time. To this end, consider equation (17) and consider the effect of a perturbation in $\kappa(\mathbf{x})$,

$$\delta\sigma(\mathbf{x}) = -\int_{\Sigma(\mathbf{x})} \delta\sqrt{\kappa} dl. \tag{23}$$

Using the definition of $\kappa(\mathbf{x})$, equation (9), I can write

$$\delta\sqrt{\kappa} = -\frac{1}{2\sqrt{\kappa}}\left[\frac{\partial\kappa}{\partial K}\delta K + \frac{\partial\kappa}{\partial S}\delta S\right] \tag{24}$$

or

$$\delta\sqrt{\kappa} = -\frac{1}{2}\left[\frac{\sqrt{\kappa}}{K}\delta K + \frac{\sqrt{\kappa}}{S}\delta S\right]. \tag{25}$$

In order to simplify the discussion, only perturbations in conductivity are considered and, after substituting the appropriate form of (25), equation (23) reduces to

$$\delta\sqrt{T} = -\int_{\Sigma}\frac{\sqrt{\kappa(s)}}{K(s)}\delta K(s)ds. \tag{26}$$

In the general case, there will be a trade-off between storage and conductivity which is difficult to resolve using arrival times alone. One possibility is to consider low frequency variations in head amplitude, variations associated with small ω. Such variations are primarily sensitive to conductivity, as is evident in equation (7). Thus, by cycling between high and low frequency data it may be possible to resolve spatial variations in both conductivity and storage.

In an actual inversion of field data one usually considers the discrete equivalent of equation (26) in which the Earth model is composed of a set of grid-blocks or cells. For example, most numerical simulators utilize a discrete grid of N cells for their computations, with each cell representing a sub-volume of the model. For each of the cells there is an average conductivity value K_j where j runs from 1 to N. Similarly, there will be a set of M residuals, $\delta\sqrt{T_i}$, in this case the square roots of the arrival times. Thus, given background values for K_j, equation (26) becomes a linear equation relating a perturbation in conductivity to a perturbation in the ith root arrival time

$$\delta\sqrt{T_i} = \sum_{j=1}^{N} M_{ij}\delta K_j \tag{27}$$

with the coefficient matrix M_{ij} given by

$$M_{ij} = -\frac{\sqrt{\kappa_j}}{K_j}\delta s_{ij} \tag{28}$$

where s_{ij} represents the length of trajectory i in cell j and κ_j and K_j are the average values of $\kappa(\mathbf{x})$ and $K(\mathbf{x})$ in cell j. The quantity that I minimize is the sum of the squares of the misfit

$$||\delta\mathbf{d} - \mathbf{M}\delta\mathbf{K}|| = \sum_{i=1}^{M}\left(\delta\sqrt{T_i} - \sum_{j=1}^{N} M_{ij}\delta K_j\right)^2 \tag{29}$$

where $||\cdot||$ denotes the L^2 norm of a vector. For a set of transient head curves I will have an associated linear system of equations which I may solve for perturbations in conductivity. Because the system of equations is extremely sparse, this is efficiently accomplished using the LSQR algorithm of [34]. Note that, due to the non-linearity of the inverse problem, I must iteratively solve the system for a series of perturbations to the initial model.

In addition to minimizing the misfit to the data I also included penalty terms to regularize (stabilize) the inversion. Inclusion of such penalty terms is standard practice for hydrological and geophysical inversions [35, 44]. I include terms which penalize large parameter deviations from a prior model and models which are not smoothly varying. The underlying motivation is that, in the absence of strong data constraints, the solution should stay close to the initial model. Also, because of the smoothing nature of hydrologic data I do not expect to resolve rapid spatial variations in flow properties. I have described this particular approach elsewhere [48] and will only outline the mechanics of the procedure here. The penalty terms most often take the form of quadratic functions on the set of models, for example model perturbation vector norm, the size of the perturbations from the prior model, is measured by

$$||\delta\mathbf{K}|| = \sum_{j=1}^{N}(\delta K_j)^2 \tag{30}$$

and model roughness, a measure of spatial variability, is given by

$$||\mathbf{L}\delta\mathbf{K}|| = \sum_{j=1}^{N}(\nabla\delta K_j)^2 \tag{31}$$

where $\mathbf{L}$ is a spatial difference operator which computes the spatial gradient of the model by differencing adjacent block values [35]. Solving the regularized inverse problem entails finding those elements of $\delta\mathbf{K}$ which minimize the sum

$$||\delta\mathbf{d} - \mathbf{M}\delta\mathbf{K}|| + ||\delta\mathbf{K}|| + ||\mathbf{L}\delta\mathbf{K}||. \tag{32}$$

The necessary equations for a minimum are an augmented linear system of the form [48]

$$\begin{pmatrix} \mathbf{M} \\ \mathbf{I} \\ \mathbf{L} \end{pmatrix} \delta\mathbf{K} = \begin{pmatrix} \delta\mathbf{d} \\ \mathbf{0} \\ \mathbf{0} \end{pmatrix}. \tag{33}$$

Amplitude Matching

The next stage of the inversion involves matching the amplitudes of the head variations. The arrival time matching produces a starting model for this part of the inversion. In my experience, the convergence of amplitude inversion algorithm is more sensitive to the starting model. Furthermore, amplitude matching is much more computationally intensive than is the arrival time inversion, and any reduction in the number of amplitude iterations is helpful. Therefore, it is a real advantage to have the arrival time result as a starting point. Due to space limitations I will not discuss amplitude matching further except to say that the sensitivities follow from the asymptotic solution (12) [56].

4 Application of the Method of Multiple Scales to Tracer Transport

In this section I use the method of multiple scales to produce a trajectory-based solution for tracer transport [52, 55]. That is, a chemical tracer is rapidly introduced into a source well while the concentration, $c(\mathbf{x}, t)$, is measured in surrounding boreholes. The governing equation is [3, 13],

$$\omega_c \frac{\partial c}{\partial t} = \nabla \cdot (\mathbf{D} \nabla c - \mathbf{U} c) \tag{34}$$

where ω_c is the kinematic porosity, $\mathbf{D}(\mathbf{x})$ is the dispersion tensor, and $\mathbf{U}$ is the Darcy velocity which is given by

$$\mathbf{U} = \mathbf{K} \cdot \nabla h \tag{35}$$

where $\mathbf{K}$ is the hydraulic conductivity tensor and $h(\mathbf{x}, t)$ is the hydraulic head [3].

In the method of multiple scales I assume that the concentration variations corresponding to the well test are much more rapid, both in space and time, than are variations in the background concentration and flow properties. Typically, the background concentration of tracer is assumed to be negligible. If I denote the scale associated with the background concentration by L and the scale of variation in the concentration associated with the tracer test by l, I am assuming that $L \gg l$. I represent the ratio l/L by the dimensionless parameter ϵ and note that $0 < \epsilon \ll 1$. I may define slow variables, in both space and time, in terms of the parameter ϵ

$$X^i = \epsilon^2 x^i \tag{36}$$

$$T = \epsilon^2 t.$$

The relationship between the slow and fast variables involves the square of ϵ. Use of these particular time and space scales is necessary if one is to balance the effects of advection or propagation against the effects of dispersion and diffusion [2]. This generalization in scaling allows one to apply asymptotic methods to dispersive and diffusive phenomena [16].

Many of the ideas developed in the method of multiple scales [2] are related to the propagation of a disturbance. Indeed, locally I shall treat the concentration variation much like a propagating wave-front. As such, I shall consider such quantities as amplitude, wave number, frequency, and phase of the front. Particularly important quantities are the amplitude and phase of the propagating concentration variation. The local phase θ is a fast or rapidly varying quantity which I define in terms of a slowly varying quantity, the phase function $\varphi(\mathbf{X}, T)$

$$\theta = \frac{\varphi(\mathbf{X}, T)}{\epsilon} = \frac{\varphi(\epsilon^2 \mathbf{x}, \epsilon^2 t)}{\epsilon}. \tag{37}$$

Formally, an asymptotic solution of equation (34) is a power series expansion in terms of the scale parameter ϵ

$$c(\mathbf{X}, T) = \sum_{n=1}^{\infty} c_n(\mathbf{X}, T, \theta)\epsilon^n. \tag{38}$$

The unknown quantities in equation (38), the functions $\theta(\mathbf{X}, T)$ and $c_n(\mathbf{X}, T, \theta)$, are determined by substituting the series into the governing equation for concentration (34) and considering terms of various orders in ϵ. I shall be particularly interested in the low-order components in ϵ, which dominate for a rapidly varying concentration. Before substituting the series (38) into equation (34) I note that the partial derivative operators in (34) may be represented in terms of derivatives with respect to the slow variables $\mathbf{X}$ and T and the phase θ

$$\begin{aligned} \frac{\partial}{\partial x^l} &= \epsilon^2 \frac{\partial}{\partial X^l} + \epsilon \frac{\partial \varphi}{\partial X^l} \frac{\partial}{\partial \theta} \\ \frac{\partial}{\partial t} &= \epsilon^2 \frac{\partial}{\partial T} + \epsilon \frac{\partial \varphi}{\partial T} \frac{\partial}{\partial \theta}. \end{aligned} \tag{39}$$

Hence, the temporal derivative in (34) takes the form

$$\frac{\partial c}{\partial t} = \epsilon^2 \frac{\partial c}{\partial T} + \epsilon \frac{\partial \varphi}{\partial T} \frac{\partial c}{\partial \theta}. \tag{40}$$

Similar expressions are obtained for the spatial derivatives contained in the gradient operator as applied to $c(\mathbf{X}, T)$, $\mathbf{U}$ and $\mathbf{D}$

$$\frac{\partial c}{\partial x^i} = \epsilon^2 \frac{\partial c}{\partial X^i} + \epsilon \frac{\partial \varphi}{\partial X^i} \frac{\partial c}{\partial \theta} \tag{41}$$

$$\frac{\partial \mathbf{U}}{\partial x^i} = \epsilon^2 \frac{\partial \mathbf{U}}{\partial X^i} \tag{42}$$

and

$$\frac{\partial \mathbf{D}}{\partial x^i} = \epsilon^2 \frac{\partial \mathbf{D}}{\partial X^i} \tag{43}$$

respectively. Higher-order spatial derivatives are obtained by successive applications of (39). If I write (34) in terms of derivatives with respect to X^i, T, and θ I have

$$\begin{aligned} \omega_c \epsilon^2 \frac{\partial c}{\partial T} + \omega_c \epsilon \frac{\partial \varphi}{\partial T}\frac{\partial c}{\partial \theta} = \\ \epsilon^4 \nabla \cdot (\mathbf{D}\nabla c) + \epsilon^4 \mathbf{D}\nabla \cdot \nabla c - \epsilon^4 \nabla \cdot \mathbf{U} c \\ + \epsilon^3 \nabla \cdot (\mathbf{D}\nabla\varphi \frac{\partial c}{\partial \theta}) + \epsilon^3 \nabla\varphi \cdot (\mathbf{D}\nabla \frac{\partial c}{\partial \theta}) \\ - \epsilon^2 \mathbf{U} \cdot \nabla c + \epsilon^2 \nabla\varphi \cdot \mathbf{D}\nabla\varphi \frac{\partial^2 c}{\partial \theta^2} \\ - \epsilon \nabla\varphi \cdot \mathbf{U} \frac{\partial c}{\partial \theta}. \end{aligned} \tag{44}$$

Terms of Order ϵ: An Equation for the Phase

Substituting the asymptotic series (38) into equation (44) I find, to first order in ϵ,

$$\omega_c \frac{\partial \varphi}{\partial T} + \nabla\varphi \cdot \mathbf{U} = 0. \tag{45}$$

I may solve this equation directly, using the method of characteristics [8]. In the method of characteristics, solutions are developed along particular trajectories, the characteristic curves. I denote the characteristic curves by $\mathbf{X}(l)$, where l is a parameter signifying position along the curve. The equations for the characteristic curves are a set of ordinary differential equations, the bi-characteristic equations,

$$\frac{d\mathbf{X}}{dl} = \mathbf{U} \tag{46}$$

$$\frac{d\sigma}{dl} = \omega_c \tag{47}$$

[8]. For a coordinate system with one axis oriented along $\mathbf{U}$ I can write (46) as

$$\frac{dr}{dl} = U \tag{48}$$

where $U = |\mathbf{U}|$ and r denotes the distance along the axis aligned with $\mathbf{U}$. Combining equations (47) and (48), I may write the 'propagation time' σ as an integral

$$\sigma = \int_\Sigma \frac{\omega_c}{U} dr \tag{49}$$

where Σ is the trajectory from the injection well to the observation well. Note that this expression for 'propagation time' is similar to that of [52, 55]. If I incorporate Darcy's law $\mathbf{U} = \mathbf{K} \cdot \nabla h$, as given by equation (35), the expression for σ, equation (49), becomes

$$\sigma = \int_\Sigma \frac{\omega_c}{|\mathbf{K} \cdot \nabla h|} dr. \tag{50}$$

Note that the integral depends on time through the time dependence of the head gradient. Also, for a general conductivity tensor **K**, the trajectory is not necessarily aligned with the head gradient. Rather, the trajectory follows the vector projection of the head gradient onto the conductivity tensor. For most tracer tests the head field has reached steady state or is maintained near steady state so that I can neglect the time variation. Equation (50) describes the evolution of a hypothetical front or discontinuity in tracer concentration which would be realized if no dispersion or diffusion occurred [18]. It is about this front that the tracer is dispersed and diffused.

Terms of Order ϵ^2: An Equation for the Amplitude

If I consider terms of order ϵ^2 in equation (44) I obtain the following expression

$$\omega_c \frac{\partial c_1}{\partial T} + \omega_c \frac{\partial \varphi}{\partial T}\frac{\partial c_2}{\partial \theta} = -\mathbf{U} \cdot \nabla c_1 + \nabla \varphi \cdot \mathbf{D} \nabla \varphi \frac{\partial^2 c_1}{\partial \theta^2} - \nabla \varphi \cdot \mathbf{U} \frac{\partial c_2}{\partial \theta} \tag{51}$$

containing amplitude coefficients c_1 and c_2 of the expansion (38). However, because the phase satisfies equation (45), terms containing c_2 drop out and I are left with

$$\omega_c \frac{\partial c_1}{\partial T} + \mathbf{U} \cdot \nabla c_1 - \nabla \sigma \cdot \mathbf{D} \nabla \sigma \frac{\partial^2 c_1}{\partial \theta^2} = 0 \tag{52}$$

an equation for the amplitude coefficient $c_1(\mathbf{X}, T, \theta)$, accurate to order ϵ^2.

I can consider the coefficient $\nabla \sigma \cdot \mathbf{D} \nabla \sigma$ in equation (52) to be a quadratic form. By a theorem of linear algebra [32], this quadratic form is equivalent to

$$Q(\mathbf{X}) = (\omega d + \alpha_l U){y_1}^2 + (\omega d + \alpha_t U){y_2}^2 + (\omega d + \alpha_t U){y_3}^2 \tag{53}$$

where

$$\mathbf{y} = \mathbf{O}^t \nabla \sigma \tag{54}$$

and **O** is an orthogonal matrix [55]. I may write (52) in terms of $Q(\mathbf{X})$

$$\omega_c \frac{\partial c_1}{\partial T} + \mathbf{U} \cdot \nabla c_1 - Q(\mathbf{X}) \frac{\partial^2 c_1}{\partial \theta^2} = 0. \tag{55}$$

Writing (55) in characteristic coordinates, given by (46) and (49), I arrive at

$$\frac{\partial c_1}{\partial \sigma} - Q(r) \frac{\partial^2 c_1}{\partial \theta^2} = 0 \tag{56}$$

where r is the position along the trajectory. I define the integral [18]

$$\tau = - \int_{\Sigma} Q(r) d\sigma \tag{57}$$

or, because $d\sigma = \omega_c/U dr$

$$\tau = -\int_{\Sigma} \frac{\omega_c Q(r)}{U} dr. \tag{58}$$

Writing equation (56) in terms of τ produces the heat equation

$$\frac{\partial c_1}{\partial \tau} + \frac{\partial^2 c_1}{\partial \theta^2} = 0. \tag{59}$$

For an impulsive point source of unit amplitude (59) has the solution [6]

$$c(\mathbf{X}, T) = \frac{1}{\sqrt{4\pi\tau}} \exp(-\frac{(T-\sigma)^2}{4\tau}). \tag{60}$$

For a general time-varying injection, I convolve equation (60) with the particular source-time function.

Solution (60) represents the contribution from a single trajectory Σ. The total tracer response is obtained by summing over all trajectorys from the injection well to the observation well. That is, equation (60) signifies the contribution from a single trajectory Σ between the boreholes. For the total response at an observation well I must integrate the responses over all contributing trajectories. If I parameterize the trajectories by a variable λ the total response is given by the integral

$$c(\mathbf{X}, T) = \int \frac{1}{\sqrt{4\pi\tau_\lambda}} \exp\left[-\frac{(T-\sigma_\lambda)^2}{4\tau_\lambda}\right] d\lambda. \tag{61}$$

The variable λ may represent the angle at which the trajectory leaves the observation well. Furthermore, it may be a vector variable in three-dimensions, signifying two angles, or an angle and position along the wellbore, necessary to specify a trajectory uniquely.

5 Application of the Method of Multiple Scales to Multiphase Flow

The method of multiple scales is used to derive a solution for two-phase flow in a heterogeneous subsurface [49, 53].

5.1 Governing Equations for Two-Phase Flow

My starting point is the set of simultaneous partial differential equations describing the flow of an aqueous (wetting) phase and a non-aqueous (non-wetting) phase [3, 13, 36]

$$\nabla \cdot \left[\frac{\rho_w K(\mathbf{x}) k_{rw}}{\mu_w} \nabla \left(P_w(\mathbf{x}, t) - \rho_w g z \right) \right] = \phi(\mathbf{x}) \frac{\partial(\rho_w S_w)}{\partial t} \tag{62}$$

$$\nabla \cdot \left[\frac{\rho_n K(\mathbf{x}) k_{rn}}{\mu_n} \nabla \left(P_n(\mathbf{x}, t) - \rho_n g z \right) \right] = \phi(\mathbf{x}) \frac{\partial(\rho_n S_n)}{\partial t}$$

where S_w and S_n denote the saturations of the aqueous and non-aqueous phases respectively. The relative permeabilities of the aqueous and non-aqueous phases, which are functions of the saturations, are represented by k_{rw} and k_{rn} while the hydraulic conductivity is given by $K(\mathbf{x})$. The respective densities are ρ_w and ρ_n, the gravitational constant is g and the porosity is $\phi(\mathbf{x})$. The pressure associated with the aqueous phase is $P_w(\mathbf{x}, t)$ while the pressure for the non-aqueous phase is $P_n(\mathbf{x}, t)$, the respective viscosities are μ_w and μ_n. The flow equations are coupled because the saturations for the two phases must sum to unity

$$S_w + S_n = 1. \tag{63}$$

Making use of the fact that $S_n = 1 - S_w$ one may derive a single equation describing the evolution of the aqueous phase saturation, which I shall denote by $S(\mathbf{x}, t)$,

$$\nabla \cdot \left[K(\mathbf{x}) F_1(S) \left\{ \Gamma \mathbf{z} - \nabla P_c(S) \right\} - F_2(S) \mathbf{q} \right] = \phi(\mathbf{x}) \frac{\partial S}{\partial t} \tag{64}$$

where $\mathbf{z}$ is the unit vector in the direction of the gravity field, $\mathbf{q}$ denotes total velocity, the sum of the velocities of the aqueous ($\mathbf{q}_w$) and non-aqueous ($\mathbf{q}_n$) phases:

$$\mathbf{q} = \mathbf{q}_n + \mathbf{q}_w, \tag{65}$$

and $F_1(S)$ and $F_2(S)$ are specified functions of saturation. In particular, $F_1(S)$ is given by

$$F_1(S) = \frac{1}{\mu_n} \frac{k_{rn}(S) k_{rw}(S)}{k_{rn}(S) + \mu_d k_{rw}(S)}. \tag{66}$$

Similarly, the function $F_2(S)$ is given by

$$F_2(S) = \frac{k_{rw}(S)}{k_{rn}(S) + \mu_d k_{rw}(S)} \tag{67}$$

and the capillary pressure is defined as

$$P_c(S) = P_n(S) - P_w(S).$$

The quantity μ_d denotes the ratio of viscosities μ_w / μ_n and Γ is the difference

$$\Gamma = g\rho_w - g\rho_n. \tag{68}$$

In general, equation (64) must be solved numerically, using a method such as integral finite differences [31, 38]. Here I derive an approximate asymptotic solution which is valid when the background saturation and flow properties are smoothly varying in a sense made more precise below. In this derivation I will neglect capillary pressure effects. A full treatment, including capillary pressure, was given in an earlier paper [49].

5.2 An Asymptotic Solution for Two-Phase Flow

There are two assumptions invoked in the derivation of an asymptotic solution for two-phase flow. First, it is assumed that the front or boundary separating the injected phase from existing fluid in the aquifer is the result of a balance between dispersive and diffusive effects and the non-linearity of two-phase propagation. Secondly, it is assumed that the background saturation and flow properties vary smoothly between known boundaries. That is, there may be discontinuities, such as layering or faults, which are modeled as boundary conditions, and smoothly varying properties between these interfaces. I represent the time and space scale of the front saturation variation by l. Similarly, the background variations are over a time and space scale L, where $L \gg l$. I represent the ratio l/L by a dimensionless parameter ϵ and require that $0 < \epsilon \ll 1$. I may define slow variables, in both space and time, in terms of the ratio ϵ:

$$\mathbf{X} = \epsilon^{\alpha}\mathbf{x} \tag{69}$$

$$T = \epsilon^{\alpha} t$$

where α is a rational number.

The quantity α is chosen such that the non-linearity balances the dispersion and dissipation [2, 43]. The formula determining α, which is based upon dimensional arguments [2], is

$$\alpha = \frac{1}{p-1} + 1, \tag{70}$$

where p is the highest order of the derivatives in the governing equations [2]. The argument is that, for a small perturbation, the most significant non-linear term is of order ϵ^2/L, assuming non-degeneracy. The effect of dispersion and dissipation is of order ϵ/L^p. For dispersion and dissipation to balance non-linearity one must have $L \sim \epsilon^{-(\alpha-1)}$. The extra ϵ enters because, in addition to the balance between non-linearity and dispersion and dissipation, I would like the front properties to vary smoothly as a function of distance. For equation (64) p equals 2 and the corresponding value of α is 2. I should note that the exact form of the scaling is not unique. Other scalings are valid and will lead to somewhat different formulations. See Korsunsky [27] for an example of two possible choices of scaling in modeling ion acoustic waves.

Much of the formalism developed in the asymptotic approach [2], is based upon concepts associated with a propagating front. For example, I shall consider properties such as the amplitude and travel time of the moving front [59]. The local travel time θ of the front is a rapidly-varying quantity which is defined in terms of a smoothly-varying function, $\varphi(\mathbf{X}, T)$

$$\theta = \frac{\varphi(\mathbf{X}, T)}{\epsilon} = \frac{\varphi(\epsilon^2\mathbf{x}, \epsilon^2 t)}{\epsilon}. \tag{71}$$

Formally, an asymptotic solution of equation (64) is a power series representation of the saturation distribution, in terms of the scale parameter ϵ

$$S(\mathbf{X},T) = s_0(\mathbf{X},T) + \sum_{n=1}^{\infty} s_n(\mathbf{X},T,\theta)\epsilon^n \tag{72}$$

where $s_0(\mathbf{X},T)$ is the background saturation distribution, which is assumed to vary smoothly in both space and time. The unknown quantities in equation (72), the functions $\theta(\mathbf{X},T)$ and $s_n(\mathbf{X},T,\theta)$, are found by substituting the series into the governing equation for saturation (64) and examining terms of various orders in ϵ. The low-order components in ϵ are of special interest for they dominate for a relatively sharp saturation front e.g. for $\epsilon = l/L \ll 1$.

Substituting the power series expansions into equation (64) and retaining terms up to order ϵ^2 results in the expression

$$\begin{aligned}
&\epsilon\nabla \cdot \left[K(\mathbf{x})\left\{F_1(s_0) + \epsilon\frac{\partial F_1}{\partial S}s_1\right\}\Gamma\mathbf{z}\right] + K(\mathbf{x})\Gamma\left\{\epsilon\frac{\partial F_1}{\partial S}\frac{\partial s_1}{\partial \theta} + \epsilon^2\frac{\partial F_1}{\partial S}\frac{\partial s_2}{\partial \theta}\right\}\nabla\varphi \cdot \mathbf{z} \\
&\quad -\epsilon K(\mathbf{x})\nabla\varphi \cdot \frac{\partial}{\partial\theta}\left\{F_1(s_0) + \epsilon\frac{\partial F_1}{\partial S}s_1\right\} \\
&\quad -\epsilon\nabla \cdot \left[\left\{F_2(s_0) + \epsilon\frac{\partial F_2}{\partial S}s_1\right\}\left\{\mathbf{q}(s_0) + \epsilon\frac{\partial \mathbf{q}}{\partial S}s_1\right\}\right] \\
&\quad -\nabla\varphi \cdot \frac{\partial}{\partial\theta}\left[\left\{F_2(s_0) + \epsilon\frac{\partial F_2}{\partial S}s_1 + \epsilon^2\frac{\partial F_2}{\partial S}s_2\right\}\right. \\
&\qquad\qquad \left.\cdot\left\{\mathbf{q}(s_0) + \epsilon\frac{\partial \mathbf{q}}{\partial S}s_1 + \epsilon^2\frac{\partial \mathbf{q}}{\partial S}s_2\right\}\right] \\
&= \epsilon\phi(\mathbf{x})\left[\frac{\partial s_0}{\partial T} + \epsilon\frac{\partial s_1}{\partial T}\right] + \phi(\mathbf{x})\frac{\partial\varphi}{\partial T}\left[\epsilon\frac{\partial s_1}{\partial\theta} + \epsilon^2\frac{\partial s_2}{\partial\theta}\right].
\end{aligned} \tag{73}$$

5.3 Terms of Order ϵ: An Equation for the Travel Time

Consideration of terms of order ϵ produces a differential equation for the travel time, φ, [49]

$$\nabla\varphi \cdot \mathbf{U} - \phi(\mathbf{x})\frac{\partial\varphi}{\partial T} = 0, \tag{74}$$

where $\mathbf{U}$ is defined as follows

$$\mathbf{U} = -K(\mathbf{x})\frac{\partial F_1}{\partial S}\Gamma\mathbf{z} + \frac{\partial F_2}{\partial S}\mathbf{q} + F_2\frac{\partial \mathbf{q}}{\partial S} \tag{75}$$

and the derivatives are evaluated with respect to the background saturation s_0, I may solve equation (74) directly, using the method of characteristics [8]. In the method of characteristics, solutions are developed along particular trajectories, the characteristic curves, which are denoted by $\mathbf{X}(l)$, where l is a parameter signifying position along the curve. The equations for the characteristic curves are a set of four ordinary differential equations

$$\frac{d\mathbf{X}}{dl} = -\mathbf{U}$$

$$\frac{dT}{dl} = \phi \tag{76}$$

[8]. For a coordinate system with one axis oriented along $\mathbf{U}$ I may write (76) as

$$\frac{dr}{dl} = -U \tag{77}$$

where $U = |\mathbf{U}|$ and r denotes the distance along the axis aligned with $\mathbf{U}$. Combining equations (76) and (77), I can write the travel time as an integral

$$T = -\int_{\Sigma} \frac{\phi}{U} dr \tag{78}$$

where Σ is the trajectory from the injection well to the observation well.

Note that, in many instances, one may associate the trajectories with streamlines used to model tracer transport and multi-phase flow [9, 10, 25]. In particular, when the vector $\mathbf{U}$, defined in equation (75), is primarily dependent on $\mathbf{q}$ the trajectories coincide with streamlines. However, if $\mathbf{q}$ depends significantly on saturation then the trajectories will deviate from streamlines. Similarly, if gravitational forces are important, the path $\mathbf{X}(l)$ will deviate from a streamline.

5.4 Terms of Order ϵ^2: An Equation for the Saturation Amplitude

Considering terms of order ϵ^2 in equation (73), leads to a nonlinear evolution equation for the saturation amplitude [49]

$$\Omega(s_0)s_1\frac{\partial s_1}{\partial \theta} + \nabla \cdot (\mathbf{U}s_1) - \phi(\mathbf{x})\frac{\partial s_1}{\partial T} = 0 \tag{79}$$

where

$$\Omega(s_0) = -2\nabla\varphi \cdot \frac{\partial \mathbf{q}}{\partial S}\frac{\partial F_2}{\partial S}.$$

I rewrite equation (79) in characteristic coordinates defined by equations (76)

$$\Omega(s_0)s_1\frac{\partial s_1}{\partial \theta} + \frac{\partial s_1}{\partial l} + \nabla \cdot \mathbf{U}s_1 = 0 \tag{80}$$

where $\nabla \cdot \mathbf{U}$ is a damping term due to gravitational forces and spatial variations in relative permeability parameters. Next, I define the variable

$$\tau = \int \Omega(s_0)dl \tag{81}$$

and rewrite equation (80) in terms of τ

$$s_1 \frac{\partial s_1}{\partial \theta} + \frac{\partial s_1}{\partial \tau} + \Phi(\tau) s_1 = 0 \tag{82}$$

where I have defined

$$\Phi(\tau) = \frac{\nabla \cdot \mathbf{U}}{\Omega(s_0)}. \tag{83}$$

Equation (82) is a generalization of an equation describing the evolution of a nonlinear wave [59]. Because equation (82) is a scalar differential equation which only depends on a single space-like variable, it may be solved efficiently using a numerical scheme such as a Total Variation Diminishing (TVD) algorithm [11].

6 Incorporating Geophysical Data

The trajectory based approach may be coupled with ray-based methods for inverting geophysical data. The efficiency of the approach is advantageous in this situation because the set of geophysical observations can be quite large.

6.1 Multi-phase Flow and Time-Lapse Seismic Data

A primary goal of almost every environmental and energy related application is to understand the distribution of flow properties within the subsurface. Important examples of such activities are: aquifer management, oil and gas production, geothermal energy development, and CO_2 sequestration. Unfortunately, our ability to image flow properties in the subsurface remains rather crude. Currently, flow properties are primarily estimated from borehole observations such as well pressures, tracer breakthroughs, and multiphase flow data. These "point" measurements are of extremely limited spatial extent, being restricted to small intervals in instrumented boreholes. Geophysical observations provide an extensive complimentary source of information. However, static geophysical data are indirectly related to flow properties in a rather complex and incompletely understood fashion. Time-lapse geophysical data, two or more geophysical surveys gathered at different times, are particularly sensitive to changes in the state of fluids in the subsurface [14, 33]. For example, in favorable situations it is possible to distinguish saturation and pressure changes from time-lapse seismic data [21, 29, 45].

Efficient Sensitivity Computation

Time-lapse seismic data can be used to infer flow properties in the sub-surface. Here I use the trajectory-based method for the modeling of multiphase flow outlined above [49, 55] to devise a highly efficient inversion scheme of time-lapse data [54]. For time-lapse data one considers the state of the reservoir at two distinct times T_0 and T_1. Quantities associated with each time are denoted by super-scripts. For illustrative purposes, consider the case in which seismic amplitude changes are primarily due to variations in saturation in both time and space. I shall relate an update in the permeability model, $\delta\mathbf{K}$, to perturbations in seismic amplitude at times T_0 and T_1, denoted by $\delta A^0(\mathbf{S}_{ij})$ and $\delta A^1(\mathbf{S}_{ij})$, respectively, where $\mathbf{S}_{ij}$ denotes a vector of saturations for the ij-th column of the reservoir, the vector index indicates the layer within the column of grid blocks. Thus, the time-lapse difference takes the form of a single summation over the perturbations in saturation within each of the k cells of the ij-th column

$$\delta A^1(\mathbf{S}_{ij}) - \delta A^0(\mathbf{S}_{ij}) = \sum_k \left(\frac{\partial A^1}{\partial S_{ijk}} \delta S^1{}_{ijk} - \frac{\partial A^0}{\partial S_{ijk}} \delta S^0{}_{ijk} \right). \tag{84}$$

The partial derivatives in (84) are computed by numerical differencing, hence any waveform computation method may be used. A rock physics model, such as Gassmann's equations [17], connects the changes in seismic moduli to changes in fluid saturations. Figure 1 displays the sensitivities associated with a reflection from the top of a reservoir model. Note that peak sensitivities are associated with saturation changes nearest to the interface, the uppermost portion of the reservoir.

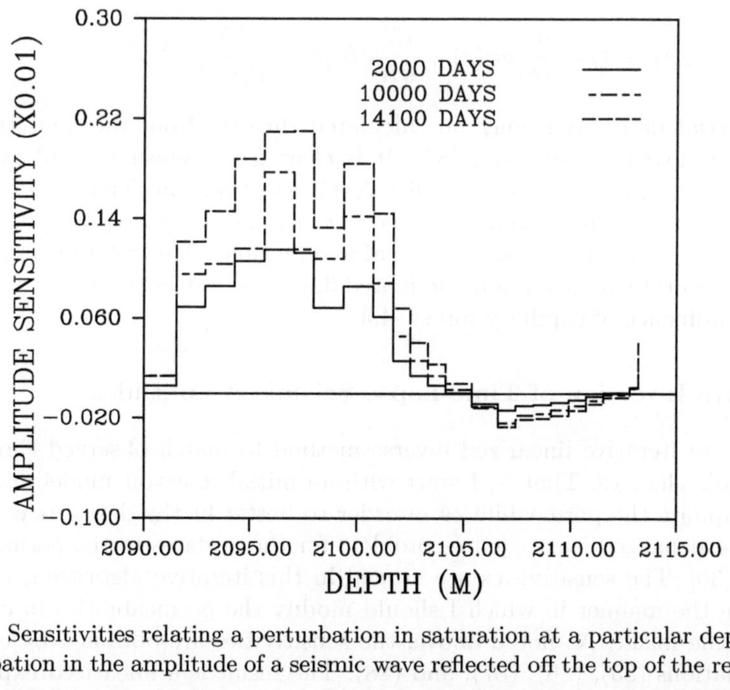

Fig. 1. Sensitivities relating a perturbation in saturation at a particular depth to a perturbation in the amplitude of a seismic wave reflected off the top of the reservoir.

The novelty of this approach involves the semi-analytic computation of $\delta {S^1}_{ijk}$ which follows from the trajectory-based modeling of multiphase flow [48, 53] and a solution of the amplitude equation (82),

$$\delta {S^1}_{ijk} = \frac{1}{T^1} S'(T^1) \int_{\mathbf{X}_{ijk}} \delta \Upsilon(\mathbf{x}) dr \tag{85}$$

where $\mathbf{X}_{ijk}$ denotes the trajectory from the ijk-th cell to a point on the initial position of the water front and the prime denotes the derivative with respect to time. The sensitivities are trajectory-based, computed as line integrals over the paths ${\mathbf{X}^1}_{ijk}$ and ${\mathbf{X}^0}_{ijk}$ in (85). The quantity $\Upsilon(r,t)$ is of the form

$$\Upsilon(r,t) = \frac{\phi}{|\mathbf{U}(r,t)|}, \tag{86}$$

the inverse of the water front velocity, as given by equation (78). Equation (86) provides a semi-analytic relationship between the water arrival time and permeability K, porosity ϕ, fluid mobility κ, and the pressure field P in the subsurface, because the water front velocity is given by [53]

$$|\mathbf{U}(r,t)| = \kappa K(r) |\nabla P(r,t)|. \tag{87}$$

Equations (84), (85), (86), and (87) relate perturbations in time-lapse seismic amplitudes to perturbations in reservoir flow properties. For a total mobility (κ) which does not vary significantly, the perturbation of $\Upsilon(r,t)$ is of the form

$$\delta \Upsilon(r,t) = \frac{\partial \Upsilon}{\partial \phi} \delta \phi(r) + \frac{\partial \Upsilon}{\partial K} \delta K(r) + \frac{\partial \Upsilon}{\partial |\nabla P|} \delta |\nabla P(r,t)|. \tag{88}$$

The partial derivatives may be calculated directly from the analytic form for $\Upsilon(r,t)$ given by (86) and (87). It is clear from equations (86) and (87) that porosity, $\phi(r)$, and permeability $K(r)$, can trade-off. That is, I can only resolve their ratio unambiguously. In order to isolate a single property, such as permeability, we must make additional assumptions. This relationship, which is most accurate for advection dominated flow, has been generalized to account for the influence of capillary forces [49].

Iterative Inversion of Time-Lapse Seismic Amplitudes

I adopt an iterative linearized inverse method to match observed time-lapse amplitude changes. That is, I start with an initial reservoir model and iteratively update the permeabilities in order to better fit the data. At each step I solve a penalized least squares problem for the updates to the permeability model [35]. The sensitivities are crucial in this iterative algorithm, for they indicate the manner in which I should modify the permeabilities in order to reduce the misfit. As noted above, the sensitivities are obtained by combining equations (85), (86), (87), and (88). The result is a linearized expression relating perturbations in time-lapse amplitude changes to perturbations in reservoir flow properties.

Because of the tradeoff between porosity and permeability, evident in (88), I cannot resolve both parameters unambiguously. Thus, I must either express one parameter in terms of the other, e.g. permeability as a function of porosity or, assume that the variation of one parameter is dominating the flow. For example, permeability can vary by many orders of magnitude and can control flow within a reservoir. The latter approach is adopted for all that follows. That is, in the application below, I shall only consider variations in the inverse of permeability, K^{-1}. Using K^{-1} normalizes the sensitivities, eliminating the K^{-2} which appears when the partial derivative of $\Upsilon(r,t)$ with respect to K is calculated [see equation (88)]. This is similar to the use of 'slowness' rather than velocity in seismic travel time tomography. Given a collection of amplitude changes, denoted by the vector $\delta\mathbf{A}^{1-0}$, by combining equations (84), (85), (86), (87) and (88) I arrive at a system of linear equations relating perturbations in time-lapse amplitude changes to perturbations in inverse permeability

$$\delta\mathbf{A}^{1-0} = \mathbf{M}\delta\mathbf{k}^{-1} \tag{89}$$

where $\delta\mathbf{k}^{-1}$ denotes a vector containing inverse permeabilities as elements, and $\mathbf{M}$ is a matrix of sensitivity coefficients. I solve equation (89) using a least-squares algorithm that is appropriate for sparse matrices [34].

The inverse problem is regularized through the addition of roughness and model norm penalty terms. Such regularization is important because, in most cases, the inverse problem is likely to be under-determined. Typically, there are many more unknown reservoir parameters than there are data. This is particularly true when the full three-dimensional inverse problem is considered. The trajectory-based sensitivities indicate that I can adjust the permeability anywhere along the trajectory in order to fit the observations. The regularization is designed to bias the updates towards smoothly varying permeability variations. That is, because I cannot resolve small scale heterogeneity, I chose to distribute the permeability updates smoothly over the entire trajectory if possible. The norm penalty term biases the result in the direction of the prior model. The roughness and norm penalty terms are quadratic forms, defined over the model space [35]. The exact penalized misfit function is of the form

$$\Pi(\mathbf{k}^{-1}) = ||\delta\mathbf{A}^{1-0} - \mathbf{M}\mathbf{k}^{-1}|| + W_n||\mathbf{k}^{-1} - \mathbf{k}^{-1}{}_0|| + W_r||\nabla\mathbf{k}^{-1}|| \tag{90}$$

where $||\cdot||$ signifies the L_2 vector norm, and W_n and W_r are the norm and roughness penalty weights, respectively [35]. The weights determine the importance of satisfying the regularization relative to fitting the observations.

Application to Time-Lapse Seismic Amplitudes from Bay Marchand, The Gulf of Mexico

The methodology has been applied to time-lapse seismic data from the Bay Marchand field in the Gulf of Mexico (Figure 2). Starting from the initial

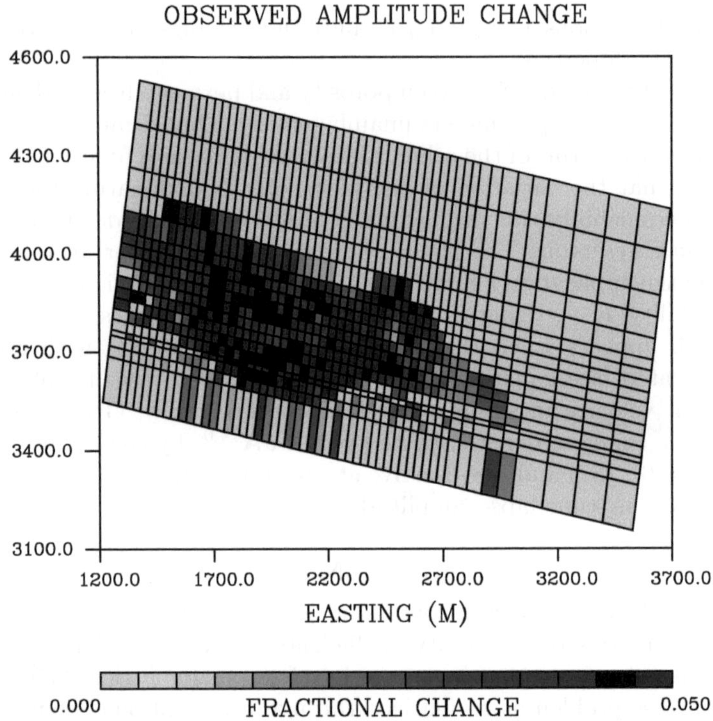

Fig. 2. Time-lapse fractional amplitude changes at Bay Marchand. The fractional amplitude change is defined by the ratio of amplitude change to the initial amplitude: $\delta\mathbf{A}^{1-0}/\mathbf{A}^0$.

permeability model shown I iteratively updated the permeabilities in order to better fit the amplitude changes. The squared misfit is reduced by 81% in twelve iterations [54]. The final model, shown in Figure 3, contains generally lower permeabilities in the central region of the reservoir. The lower permeabilities are required to slow down the arrival of the water in order to produce the largest changes within the time interval between the two seismic surveys.

6.2 Transient Head and Tilt Data

In this sub-section I discuss a method for inferring flow properties from borehole head and surface deformation data. In essence, the methodology is a blending of techniques presented in Vasco et al. [56] and [57], combining some of the advantages of each method. First, the surface deformation data are used, along with downhole pressure measurements, to infer head variations within the aquifer as a function of time. Second, the temporal variations in head throughout the aquifer are used to define a head 'arrival time'. As shown in Vasco et al. [56] such arrival times may be related directly to flow properties. Using this approach, I retain the quasi-linearity, efficiency, and robustness of

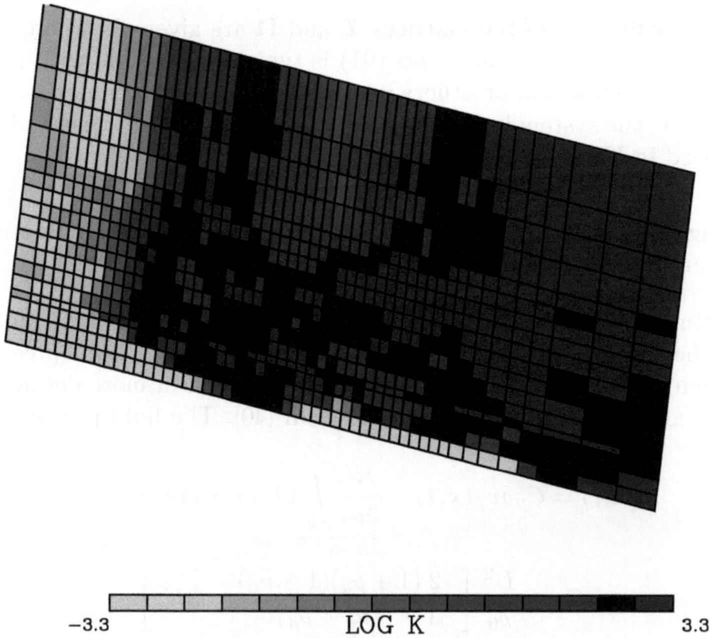

Fig. 3. Permeability distribution resulting from an inversion of the time-lapse amplitude changes.

head arrival time inversion. In developing a coupled surface deformation and borehole pressure inversion for flow properties I can utilize surface observations. Such observations are generally much less expensive and less intrusive than are borehole measurements.

Estimation of Hydraulic Conductivity from Borehole Pressure and Surface Deformation Data

Estimation of volume change The first step is to use the surface deformation data to infer variations in hydraulic head or pressure. This aspect has already been discussed thoroughly [57]. Essentially, this step requires the solution of a set of linear equations. This is a linear inverse problem for a discrete representation of reservoir volume change. Using linear poroelasticity one may relate changes in pressure or hydraulic head to fluid volume changes in a permeable region of the Earth. Thus, I may conduct a coupled inversion of pressure or head measurements and surface deformation observations [57]. I will only state the final result, a system of linear equations relating volume changes within each grid block of a reservoir model $\delta \mathbf{v}_f(t)$ to pressure or head measurements $\mathbf{p}(t)$ and surface deformation observations $\mathbf{u}(t)$

$$\begin{bmatrix} \mathbf{u}(t) \\ \mathbf{p}(t) \end{bmatrix} = \begin{bmatrix} \boldsymbol{\Upsilon} \\ \boldsymbol{\Pi} \end{bmatrix} \delta \mathbf{v}_f(t) \tag{91}$$

where the coefficients of the matrices $\boldsymbol{\Upsilon}$ and $\boldsymbol{\Pi}$ are given elsewhere [50]. The solution of the system of equations (91) is typically unstable with respect to perturbations, numerical or otherwise, and some form of regularization is required. Thus, the system is solved using a regularized least squares algorithm, as discussed in Vasco et al. [56, 57].

Mapping Estimates of Volume Change into Changes in Head and Overburden Stress

Once estimates of the volume change within the reservoir are available, by solving the system of equations (91), I map them into fluid pressure and overburden stress changes. The mapping is described in more detail in Vasco et al. [57] and is based on the work of Segall [40]. The fluid pressure is given by

$$p(\mathbf{x},t) = C_\nu \delta v_f(\mathbf{x},t) - \frac{B^2}{3\rho_0}\int_V G_{ii}(\mathbf{x},\mathbf{y})\delta v_f(\mathbf{y},t)dV \tag{92}$$

where

$$C_\nu = \frac{B^2}{\rho_0}\left[\mu\frac{2}{3}\frac{(1+\nu_d)(1+\nu_u)}{(\nu_u-\nu_d)} - 3K_u\right], \tag{93}$$

ν_u and ν_d are the undrained and drained Poisson's ratios [39]. The pore pressure estimates are converted to head using the definition

$$h = \frac{p}{\rho g} + z. \tag{94}$$

Note that all of the mappings are linear, as is the inverse problem for volume change, equation (91).

Determination of Flow Properties from Head Estimates

Given estimates of head and overburden pressure changes, I may now solve for flow properties within the reservoir. The procedure for estimating flow properties is based upon an asymptotic solution of the equation for head, equation (12), as described in Vasco et al. [56] and Vasco and Finsterle [55]. The asymptotic solution shares many properties with ray methods which are used in a wide range of imaging algorithms. In particular, the solution is defined on a trajectory, or curve $\mathbf{X}(s)$, through the model, similar to an optical ray. The scalar variable s is a measure of distance along the trajectory.

Due to the coupling between pressure changes within the overburden and head changes within the reservoir, the results of Vasco et al. [56] must be modified somewhat. The additional complication is due to the time derivative of overburden pressure as a source term [50]. Because of this additional term, the source is no longer a step-function, restricted to the wellbore, as was assumed in Vasco et al. [56]. Rather, the bulk pressure term acts as a spatially-distributed source with a time-varying magnitude. In an earlier paper [50] I show how the head variations are modified, or corrected, to account for time-varying stresses in the overburden. One can then treat the corrected head variation using techniques from Vasco et al. [56]. Typically, the corrections will be small but they may be observable in particular circumstances.

For a step-function source, the corrected asymptotic solution for the time derivative of head retains the form

$$\dot{h}_c(\mathbf{X}, t) = Q_0 A_0(\mathbf{X}) \frac{\sigma(\mathbf{X})}{2\sqrt{\pi t^3}} e^{-\sigma^2(\mathbf{X})/4t} \tag{95}$$

[50] where the dot denotes differentiation with respect to time, A_0 is the amplitude associated with the trajectory, Q_0 is the flow rate, and σ is the 'phase', which is defined as

$$\sigma = \sqrt{6T(\mathbf{X})} = \int_{\mathbf{X}} \sqrt{\kappa(s)} ds \tag{96}$$

where $T(\mathbf{X})$ is the travel time from the pumping well to the observation point, and $\kappa(s)$ is the ratio

$$\kappa(s) = \frac{S}{K} \tag{97}$$

where S is the specific storage in equation (6) and K is the hydraulic conductivity. It was shown in Vasco et al. [56] and Vasco and Finsterle [55], following Virieux et al. [58] that, for a constant rate pump test, the quantity $T(\mathbf{X})$ is the time at which the derivative of the head, $\dot{h}_c$, is a maximum. As noted above and discussed in [55], it is possible to compute $T(\mathbf{X})$, and correspondingly $\sigma(\mathbf{X})$, using a numerical simulator.

The arrival times of the drawdown are matched using an iterative, linearized least squares approach [55, 56] which was introducted above [equations (23)-(33)]. Thus, one can derive a linearized relationship between the perturbed arrival time and the perturbation in conductivity

$$\delta\sqrt{T} = -\int_{\Sigma} \frac{\sqrt{\kappa(s)}}{K(s)} \delta K(s) ds. \tag{98}$$

The path of integration is the trajectory Σ, extending from the pumping well to the observation point. This curve is computed using the flow properties of the initial model, $K_0(\mathbf{X})$. The integral in (98) is discretized by considering the path length in each grid block and writing (98) as a sum over the segments in each cell. The result is a linear constraint, relating the weighted sum of conductivity perturbations in the grid blocks traversed by the cells to the total travel time along the path.

There will be a set of arrival times, one for each of the grid blocks in the reservoir model. The collection of linear constraints provided by the arrival times may be written in matrix-vector form as

$$\delta\mathbf{T} = \mathbf{M}\delta\mathbf{K} \tag{99}$$

where $\mathbf{M}$ is the matrix with coefficients determined by the discretization of the integral (98), $\delta\mathbf{K}$ is a vector of conductivity updates for each grid block, and $\delta\mathbf{T}$ is the vector of arrival time observations. The linear system (99) is generally sparse. The non-zero coefficients of $\mathbf{G}$ correspond to the grid blocks which are intersected by the trajectories. As discussed above, equation (99) is augmented with additional equations corresponding to penalty terms used to regularize the inverse problem. The conductivity updates $\delta\mathbf{K}$ are then added to the background model $\mathbf{K}_0$. Then a numerical simulator is used to compute an updated drawdown history in the new model. The head values are corrected, as noted above, and discussed in [50], and new trajectories and residuals are calculated. A new set of equations (99) is constructed and solved. The entire process is repeated until the misfit converges to an acceptable value.

A Pump Test at the Raymond Field Site

The Raymond Quarry test site, in the foothills of the Sierra Nevada, served as a natural laboratory for evaluating geophysical and hydrological techniques for mapping fractures [23]. Because the major pathway for flow in many rock types is through fractures, it is important to develop techniques for characterizing flow and transport in fractures. Even if the geometry of a fracture is known, say from intersections with various boreholes, the conductivity within the fracture plane may vary by orders of magnitude. Therefore, it is essential that one develop techniques to estimate fracture flow properties.

An extended production test, of approximately 2 hours duration, was conducted on the morning of the 15th of August, 1995. Water was pumped from well 00, the well located at the apex of an inverted V pattern of wells, at a rate of 0.4 L/s. Vasco et al. [57] analyzed the tilt data that resulted from that test. In particular, Vasco et al. [57] conducted a series of coupled inversions of the head and tilt data in order to estimate the distribution of volume change within the upper fracture zone as a function of time. The fracture zone was sub-divided into a 15 by 15 grid of cells, each of which was allowed to undergo a distinct fractional volume change. A regularized, least squares inversion of the tilt and head data provided estimates of the 225 unknown fractional volume changes [57]. Figure 4 shows the results of those coupled inversions of head and tilt, for the first six minutes of pumping. These estimates are the starting point for the current analysis. Specifically, I map the fractional volume changes into fluid and overburden pressure variations using equation (92). Next, I map the pressure changes into head and estimate the time derivatives using a spline interpolation algorithm. The head derivatives are corrected for the effects of overburden pressure changes, as described in [50]. The corrections for overburden pressure changes proved to be small. The resulting head slope variations for 15 grid blocks in the fracture zone model are shown in Figure 5. Note the variation in amplitude and the arrival time of the peak slope for each grid block. The estimated arrival times for all the grid blocks are shown in Figure 6. I do not compute an arrival time esti-

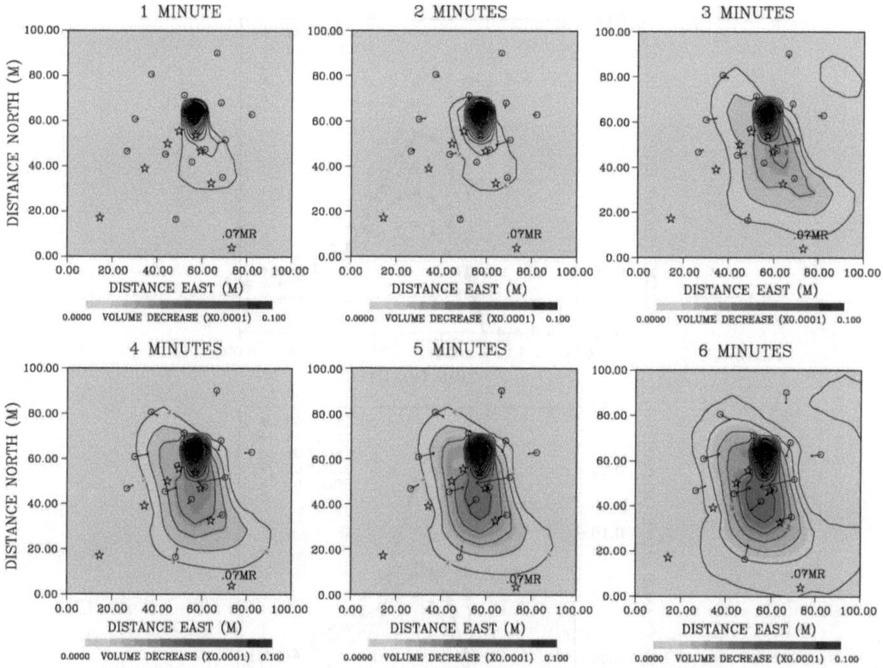

Fig. 4. Volume change within the fracture zone after the start of pumping

mate if the head variation is less than 10% of the peak variation. In Figure 6, the arrival times of those cells are set to a maximum value of five minutes for the purposes of plotting, but the arrival times are not used in the inversion for conductivity. In Figure 6 one observes the rapid propagation of drawdown to the south-southeast of the pumping well.

The arrival time estimates for the active grid blocks form the data I use to infer conductivity variations within the fracture zone. The relationship between arrival time variations and conductivity updates to the fracture zone model is given by equation (98). In all there were 50 arrival times, corresponding to the active cells in Figure 6. The 50 equations defining the data constraints, form a linear system for the updates to the conductivity model, equation (99). These equations are augmented with equations penalizing the model roughness [55, 56]. The resulting sparse system of linear equations is solved using the LSQR algorithm [34].

Because the inverse problem is non-linear I iteratively refine the model in a sequence of linearized steps. That is, I solve the system of equations (99) for conductivity perturbations. The conductivity model is updated and a new flow simulation is run. New trajectories and residuals form the basis for another inversion step and another update. The process is repeated until satisfactory convergence is achieved. After 15 or so iterations the misfit is reduced by over two orders of magnitude. Furthermore, the misfit reduction for each subsequent step is rather small. Therefore, after 25 iterations the

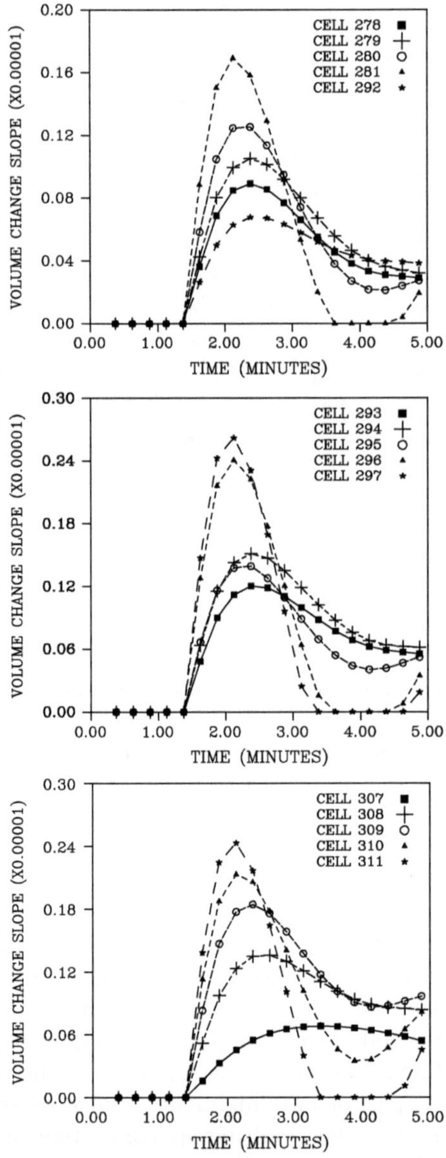

Fig. 5. The interpolated head derivative for 15 grid-blocks

inversion is concluded. The initial fits are highly scattered but generally the predicted arrival time is much too large, suggesting permeabilities which are too low. After the inversion, the predicted arrival times are much closer to the observed values.

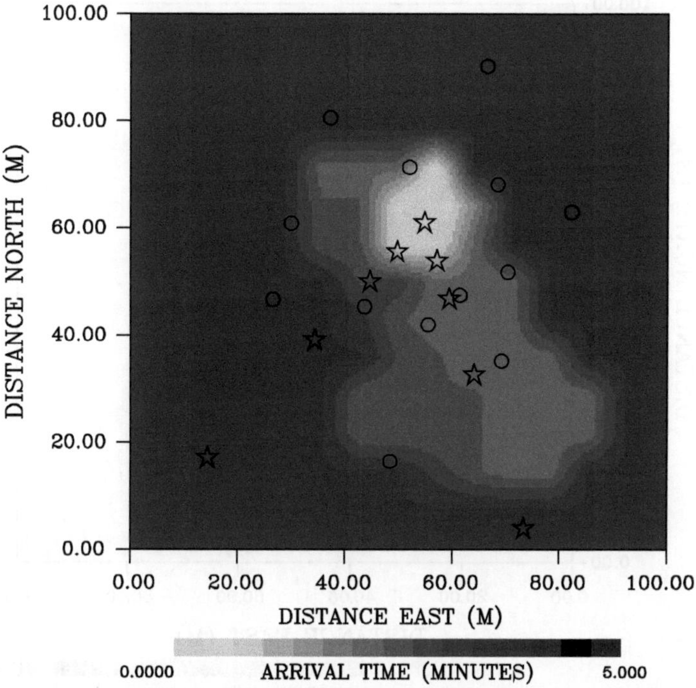

Fig. 6. Estimated arrival times for the array of grid-blocks. The open stars denote well locations in which pressure sensors were deployed. The open circles denote the locations of tiltmeters used in the pumping experiment.

The final conductivity model is shown in Figure 7. The main feature is an elongated high conductivity zone which extends to the south of the pumping well, sub-parallel to the eastern edge of the array of wells. The result agrees with previous estimates [57], which are based on a completely different approach. The main difference is the absence of high conductivity feature to the northwest of the pumping wells [57]. Perhaps this anomaly is not recovered due to the lack of sufficient tilt data to the north.

7 Discussion and Conclusions

In this chapter I have shown how asymptotic techniques can be used to develop semi-analytic solutions for modeling flow and transport. The asymptotic approach results in solutions which are defined on trajectories or paths through the medium. There are several advantages associated with these types of solutions, particularly when considering the inverse problem of infering flow properties from sets of observations. For example, the closed form solutions may be used to derive semi-analytic expressions for model parameter sensitivities. Furthermore, because the solutions are defined along trajectories

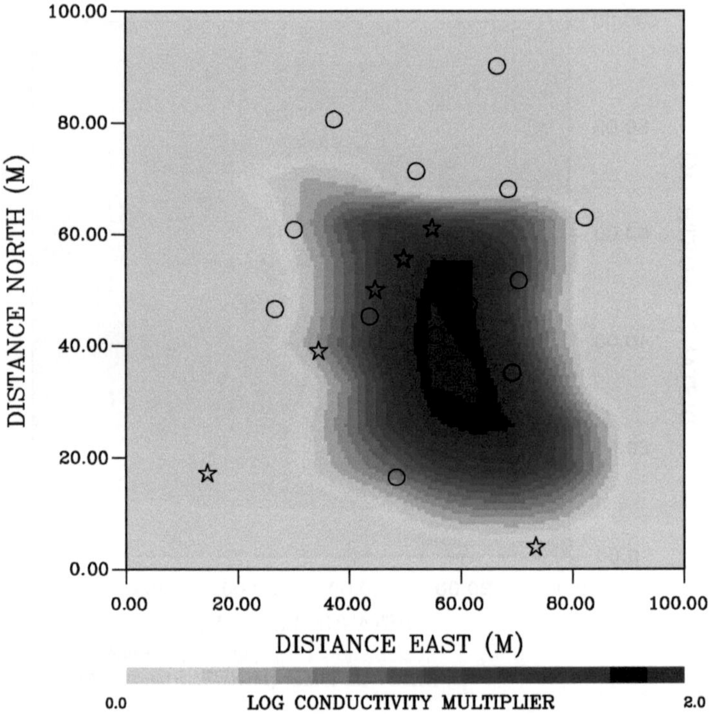

Fig. 7. Conductivity estimates resulting from an inversion of the arrival times in Figure 6. The quantity 'log conductivity multiplier' denotes a scalar multiplier of the logarithm of the conductivity.

through the model, the order of computation and memory usage scales as roughly the number of grid blocks intersected by the model. For an n_x by n_y by n_z grid of cells, the computation is of the order of $\sqrt{{n_x}^2 + {n_y}^2 + {n_z}^2}$. For a fully numerical method, such as finite differences, the computation may scale as $n_x \times n_y \times n_z$. Another advantage of the asymptotic or trajectory-based approach is that the problem de-couples into an arrival time matching problem and an amplitude matching problem. The arrival time matching problem is quasi-linear and generally involves much less computation than does amplitude matching [7]. Thus, an arrival time inversion can provide a useful starting model for amplitude matching [52]. The computational advantages should prove particularly helpful when utilizing geophysical data sets which can be quite large. In addition, the techniques should prove useful when considering more complicated physics, such as coupled flow and reactive transport. These are the areas where future work should prove fruitful and may provide an important complement to fully numerical techniques.

References

1. K. Aki and P. G. Richards. *Quantitative Seismology.* Freeman and Sons, 1980.
2. A. M. Anile, J. K. Hunter, P. Pantano, and G. Russo. *Ray Methods for Nonlinear Waves in Fluids and Plasmas.* Longman Scientific and Technical, 1993.
3. J. Bear. *Dynamics of Fluids in Porous Media.* Dover Publications, 1972.
4. R. N. Bracewell. *The Fourier Transform and Its Applications.* McGraw-Hill, New York, 1978.
5. J. Carrera. State of the art of the inverse problem applied to the flow and solute transport problems. In *Groundwater Flow and Quality Modeling*, pages 549–585. NATO ASI Ser., 1987.
6. H. S. Carslaw and J. C. Jaeger. *Conduction of Heat in Solids.* Oxford University Press, Oxford, 1959.
7. H. Cheng, A. Datta-Gupta, and Z. He. A comparison of travel-time and amplitude matching for field-scale production data integration: Sensitivity, non-linearity, and practical implications. *SPE Annual Meeting*, 84570:1–15, 2003.
8. R. Courant and D. Hilbert. *Methods of Mathematical Physics.* Interscience, New York, 1962.
9. M. J. Crane and M. J. Blunt. Streamline-based simulation of solute transport. *Water Resour. Res.*, 35:3061–3078, 1999.
10. A. Datta-Gupta and M. J. King. A semianalytic approach to tracer flow modeling in heterogeneous permeable media. *Advan. Water Resour.*, 18:9–24, 1995.
11. A. Datta-Gupta, L. W. Lake, G. A. Pope, K. Sepehrnoori, and M. J. King. High-resolution monotonic schemes for reservoir fluid flow simulation. *In Situ*, 15:289–317, 1991.
12. A. Datta-Gupta, D. W. Vasco, J. C. S. Long, P. S. D'Onfro, and W. D. Rizer. Detailed characterization of a fractured limestone formation by use of stochastic inverse approaches. *SPE Form. Eval.*, 10:133–140, 1995.
13. G. de Marsily. *Quantitative Hydrogeology.* Academic Press, San Diego, 1986.
14. S. N. Domenico. Effect of water saturation on seismic reflectivity of sand reservoirs encased in shale. *Geophysics*, 39:759–769, 1974.
15. F. G. Friedlander and J. B. Keller. Asymptotic expansions of solutions of $(\nabla^2 + k^2)u = 0$. *Comm. Pure Appl. Math.*, 8:387, 1955.
16. C. S. Gardner and G. K. Morikawa. Similarity in the asymptotic behavior of collision-free hydromagnetic waves and water waves. *Courant Inst. Math. Sci. Rep.*, 90:9082, 1960.
17. F. Gassmann. Elastic waves through a packing of spheres. *Geophysics*, 16:673–685, 1951.
18. L. W. Gelhar and M. A. Collins. General analysis of longitudinal dispersion in nonuniform flow. *Water Resour. Res.*, pages 1511–1521, 1971.
19. R. E. Grundy, C. J. van Duijn, and Dawson C. N. Asymptotic profiles with finite mass in one-dimensional contaminant transport through porous media: The fast reaction case. *Q. J. Mech. Appl. Math.*, 47:69–106, 1994.
20. Z. He, A. Datta-Gupta, and D. W. Vasco. Rapid inverse modeling of pressure interference tests using trajectory-based traveltime and amplitude sensitivities. *Water Resources Research*, 42:1–15, 2006.
21. G. M. Hoversten, R. Gritto, J. Washbourne, and T. Daley. Pressure and fluid saturation prediction in a multicomponent reservoir using combined seismic and electromagnetic imaging. *Geophysics*, 68:1580–1591, 2003.
22. U. Jaekel, A. Georgescu, and H. Vereecken. Asymptotic analysis of nonlinear equilibrium solute transport in porous media. *Water Resour. Res.*, 32:3093–3098, 1996.
23. K. Karasaki, B. Freifeld, A. Cohen, K. Grossenbacher, P. Cook, and D. Vasco. A multidisciplinary fractured rock characterization study at the Raymond field site, Raymond, California. *J. Hydrol.*, 236:17–34, 2000.

24. J. B. Keller and R. M. Lewis. Asymptotic methods for partial differential equations: The reduced wave equation, and Maxwell's equations. In J. P. Keller, D. W. McLaughlin, and G. C. Papanicolaou, editors, *Surveys in Applied Mathematics*, volume 1, chapter 1. Plenum Press, 1995.
25. M. J. King and A. Datta-Gupta. Streamline simulation: A current perspective. *In Situ*, 22:91–140, 1998.
26. M. Kline and I. W. Kay. *Electromagnetic Theory and Geometrical Optics*. Robert E. Krieger Publishing, Huntington, NY, 1979.
27. S. Korsunsky. *Nonlinear Waves in Dispersive and Dissipative Systems with Coupled Fields*. Addison Wesley Longman Ltd., Essex, 1997.
28. Y. A. Kravtsov and Y. I. Orlov. *Geometrical Optics of Inhomogeneous Media*. Springer-Verlag, 1990.
29. M. Landro. Discrimination between pressure and fluid saturation changes from time-lapse seismic data. *Geophysics*, 66:836–844, 2001.
30. R. K. Luneburg. *Mathematical Theory of Optics*. University of California Press, Berkeley, 1966.
31. T. N. Narasimhan and P. A. Witherspoon. An integrated finite difference method for analyzing fluid flow in porous media. *Water Resour. Res.*, 12:57–64, 1976.
32. B. Noble and J. W. Daniel. *Applied Linear Algebra*. Prentice-Hall, Englewood Cliffs, NJ, 1977.
33. A. Nur. Four-dimensional seismology and (true) direct detection of hydrocarbon: The petrophysical basis. *The Leading Edge*, 8:30–36, 1989.
34. C. C. Paige and M. A. Saunders. LSQR: An algorithm for sparse linear equations and sparse linear systems. *ACM Trans. Math. Software*, 8:195–209, 1982.
35. R. L. Parker. *Geophysical Inverse Theory*. Princeton University Press, Princeton, 1994.
36. D. W. Peaceman. *Fundamentals of Numerical Reservoir Simulation*. Elsevier Scientific Publishing, 1977.
37. W. H. Press, S. A. Teukolsky, W. T. Vetterling, and B. P. Flannery. *Numerical Recipes*. Cambridge University Press, Cambridge, 1992.
38. K. Pruess, C. Oldenburg, and G. Moridis. Tough2 user's guide, version 2.0. *LBNL Report, Berkeley*, 43134, 1999.
39. J. R. Rice and M. P. Cleary. Some basic stress diffusion solutions for fluid-saturated elastic porous media with compressible constituents. *Rev. Geophys. Space. Phys.*, 14:227–241, 1976.
40. P. Segall. Stress and subsidence resulting from subsurface fluid withdrawal in the epicentral region of the 1983 Coalinga earthquake. *J. Geophys. Res.*, 90:6801–6816, 1985.
41. J. A. Sethian. *Level Set and Fast Marching Methods*. Cambridge University Press, Cambridge, 1999.
42. N.-Z. Sun and W.-G. Yeh. Coupled inverse problems in groundwater modeling, 1, sensitivity analysis and parameter identification. *Water Resour. Res.*, 26:2507–2525, 1990.
43. T. Taniuti and Nishihara K. *Nonlinear Waves*. Pitman Publishing Inc., Marshfield, MA, 1983.
44. A. Tarantola. *Inverse Problem Theory: Methods for Data Fitting and Model Parameter Estimation*. Elsevier Sci., 1987.
45. A. Tura and D. E. Lumley. Estimating pressure and saturation changes from time-lapse avo data. *61st Ann. Conf. Eur. Assoc. Geosci. Eng.*, Extended Abstracts:1–38, 1999.
46. S. E. van der Zee. Analytical traveling wave solutions for transport with nonlinear and nonequilibrium adsorption. *Water Resour. Res.*, 26:2563–2578, 1990.

47. C. J. van Duijn and P. Knabner. Traveling waves in the transport of reactive solutes through porous media: Adsorption and binary ion exchange, 1. *Trans. Porous. Media*, 8:167–194, 1992.
48. D. W. Vasco, S. Yoon, and A. Datta-Gupta. Integrating dynamic data into high-resolution reservoir models using streamline-based analytic sensitivity coefficients. *SPE Journal*, 4:389–399, 1999.
49. D. W. Vasco. An asymptotic solution for two-phase flow in the presence of capillary forces. *Water Resources Research*, 40:W12407, 2004.
50. D. W. Vasco. Estimation of flow properties using surface deformation and head data: A trajectory-based approach. *Water Resources Research*, 40:1–14, 2004.
51. D. W. Vasco. Trajectory-based modeling of broadband electromagnetic wavefields. *Geophysical Journal International*, 168:949–963, 2007.
52. D. W. Vasco and A. Datta-Gupta. Asymptotic solutions for solute transport: A formalism for tracer tomography. *Water Resources Research*, 35:1–16, 1999.
53. D. W. Vasco and A. Datta-Gupta. Asymptotics, saturation fronts, and high resolution reservoir characterization. *Transport in Porous Media*, 42:315–350, 2001.
54. D. W. Vasco, A. Datta-Gupta, R. Behrens, P. Condon, and J. Rickett. Seismic imaging of reservoir flow properties: Time-lapse amplitude changes. *Geophsics*, 69:1425–1442, 2004.
55. D. W. Vasco and S. Finsterle. Numerical trajectory calculations for the efficient inversion of transient flow and tracer observations. *Water Resources Research*, 40:W01507, 2004.
56. D. W. Vasco, K. Karasaki, and H. Keers. Estimation of reservoir properties using transient pressure: An asymptotic approach. *Water Resources Research*, 36:3447–3465, 2000.
57. D. W. Vasco, K. Karasaki, and K. Kishida. A coupled inversion of pressure and surface displacement. *Water Resources Research*, 37:3071–3089, 2001.
58. J. Virieux, C. Flores-Luna, and D. Gibert. Asymptotic theory for diffusive electromagnetic imaging. *Geophys. J. Int.*, 119:857–868, 1994.
59. G. B. Whitham. *Linear and Nonlinear Waves.* Wiley, New York, 1974.
60. W.-G. Yeh. Review of parameter identification procedures in groundwater hydrology: The inverse problem. *Water Resour. Res.*, 22:95–108, 1986.

The Role of Streamline Models for Dynamic Data Assimilation in Petroleum Engineering and Hydrogeology

Akhil Datta-Gupta, Deepak Devegowda, Dayo Oyerinde, and Hao Cheng

Texas A&M University

1 Introduction

Geological models derived from static data alone often fail to reproduce the dynamic response from the aquifers, for example transient pressure or tracer response or from petroleum reservoirs, for example multiphase production history such as water cut or gas-oil ratio. Reconciling geologic models to the dynamic response of the reservoir is critical for subsurface characterization and building reliable reservoir performance models. Available information on subsurface heterogeneity can be broadly categorized into two major types: static and dynamic. Static data are time-invariant direct or indirect measurements of reservoir properties, such as cores, well logs, and 3-D seismic data. With recent advances in reservoir characterization, these data can now be integrated efficiently into coherent 3-D reservoir descriptions (Dubrule, 1998). Dynamic data are the time dependent measurements of flow responses such as pressure, flow rate, fractional flow and, with the use of 4-D seismic, time-lapse saturation and pressure. Integration of dynamic data generally leads to an inverse problem and requires solution of the flow equations several times using an iterative procedure (Hyndman et al., 1994; Kitanidis, 1995; Mclaughlin and Townley, 1996; Medina and Carrera, 1996; Anderman and Hill, 1999; Vasco and Datta-Gupta, 1999; Yeh and Liu, 2000; Oliver et al., 2001). The process is commonly referred to as "history matching" and is usually the most tedious and time-consuming aspect of subsurface flow and transport simulation study.

Streamline models offer some unique advantages in history matching (Vasco et al., 1999; Datta-Gupta et al., 2002; He et al., 2006). In this chapter we will explore the properties of streamline models that make them particularly attractive for the integration of production data into high resolution geologic models. The streamline approach has provided an extremely efficient means for computing parameter sensitivities which define the relationship between subsurface parameters and production response (Datta-Gupta et al., 2001). The streamline-based sensitivity computation has been extended to include gravity, changing field conditions and more recently to fractured

A. Datta-Gupta et al.: *The Role of Streamline Models for Dynamic Data Assimilation in Petroleum Engineering and Hydrogeology*, Studies in Computational Intelligence (SCI) **79**, 105–136 (2008)
www.springerlink.com

reservoirs (He et al., 2002; Al Harbi et al., 2005). With the streamline method, the sensitivities can be computed analytically using a single flow simulation. Because the sensitivity calculations involve evaluation of 1-D integrals along streamlines, the method scales very well with respect to model size and is thus, well suited for production data integration into highly detailed geologic models. Most of the previous applications of streamline-based history matching have been limited to two-phase water-oil flow under incompressible or slightly compressible conditions. This is partly because of the lack of availability of a rigorous compressible formulation for streamline simulation and the associated analytic sensitivity computations for compressible and three-phase flow.

In this chapter, we first discuss the generalization of streamline models to compressible flow using a rigorous formulation while retaining most its computational advantages. Our formulation fully accounts for the compressibility effects on the flow and transport calculations but requires only minor modifications to existing streamline models. This is accomplished by introducing an 'effective density' of the total fluids along the streamlines (Cheng et al., 2006; Osako and Datta-Gupta, 2007). This density captures the changes in the fluid volume with pressure and can be conveniently and efficiently traced along streamlines. Our approach preserves the 1-D nature of the saturation calculations and all the associated advantages of the streamline models. Next, we generalize the streamline-based history matching to compressible and three-phase flow including water, oil and gas phases. Specifically, we formulate analytic expressions for the sensitivity of water cut and gas-oil ratio to reservoir properties, in particular, permeability. Again, the sensitivities are simple 1-D integrals along streamlines and can be computed using a single forward simulation. For history matching, we use a generalized travel time inversion that has been successfully applied to many field cases involving two-phase flow. Our work here generalizes the approach to three phase flow by incorporating matching of both water cut and gas-oil ratio. Finally, we briefly discuss the role of streamline models in uncertainty assessment via a multistage sampling algorithm.

2 Streamline Simulation: Background and Generalization to Compressible Flow

Streamline models have seen resurgence in interest because of their ability to efficiently simulate fluid flow and transport through highly detailed geologic models (Datta-Gupta, 2000). Streamline simulators approximate three-dimensional fluid flow calculations by a sum of one-dimensional solutions along streamlines. The choice of streamline directions for the one dimensional calculations makes the approach extremely effective for modeling convection dominated flows in the reservoir. This is typically the case when heterogeneity dominates the flow behavior. The streamlines are generally distributed in space with higher resolution than the underlying spatial grid, thus providing excellent transverse resolution in regions of fast flow. Transport calculations along streamlines are decoupled from the underlying grid and can be carried out with little or no intrinsic time-step limitations (Bratvedt et al., 1996; King and Datta-Gupta, 1998).

2.1 Background

We will first review the current streamline formulation before discussing its extensions to compressible flow. At a fundamental level, all streamline techniques are based upon a coordinate transformation from physical space to a coordinate system following the flow directions. This transformation is based upon the bi-streamfunctions and an additional time of flight coordinate. Following Bear (1973) we define a streamline which is everywhere tangential to the velocity field, by introducing the bi-streamfunctions Ψ and χ,

$$\vec{u} = \nabla\psi \times \nabla\chi \tag{1}$$

Note that the incompressibility assumption is implicit in this representation because of the vector identity,

$$\nabla \bullet (\nabla\psi \times \nabla\chi) = 0 \tag{2}$$

A streamline is defined by the intersection of a constant value for Ψ with a constant value for χ. An important concept in streamline simulation is the 'time of flight', τ, which is defined simply as the travel time of a neutral tracer along the streamlines (Datta-Gupta and King, 1995),

$$\tau(x, y, z) = \int \frac{\phi ds}{|\vec{u}|} \tag{3}$$

Or, in a differential form as follows

$$\vec{u} \cdot \nabla\tau = \phi \tag{4}$$

Streamline techniques are based upon a coordinate transformation from the physical space to the time of flight coordinate where all the streamlines can be treated as straight lines of varying lengths. This coordinate transformation is greatly facilitated by the fact that the Jacobian of the transformation takes an extraordinarily simple form,

$$\left\| \frac{\partial(\tau, \psi, \chi)}{\partial(x, y, z)} \right\| = (\nabla\psi \times \nabla\chi) \bullet \nabla\tau = \vec{u} \bullet \nabla\tau = \phi \tag{5}$$

We now have the following relationship between the physical space and the time of flight coordinates,

$$\phi dx\ dy\ dz = d\tau d\psi\ d\chi \tag{6}$$

Clearly, the coordinate transformation provides a simple representation for the pore volume in streamline coordinates; this is an essential feature to maintain the material balance.

Spatial gradients along streamlines assume a very simple form in the time of flight coordinates. Using the (τ, ψ, χ) coordinates the gradient operator can be expressed as,

$$\nabla = (\nabla \tau) \frac{\partial}{\partial \tau} + (\nabla \psi) \frac{\partial}{\partial \psi} + (\nabla \chi) \frac{\partial}{\partial \chi} \tag{7}$$

Because $\vec{u}$ is orthogonal to both $\nabla \psi$ and $\nabla \chi$,

$$\vec{u} \bullet \nabla = \phi \frac{\partial}{\partial \tau} \tag{8}$$

Convection-driven flow now reduces to one dimensional spatial gradients along streamlines. For instance, let us consider the conservation equation for single phase tracer transport under incompressible flow conditions,

$$\phi \frac{\partial C}{\partial t} + \vec{u} \bullet \nabla C = 0 \tag{9}$$

This expression can be transformed into the τ coordinate using Eq. (8),

$$\frac{\partial C}{\partial t} + \frac{\partial C}{\partial \tau} = 0 \tag{10}$$

After this coordinate transformation, the three dimensional fluid flow has been decomposed into a series of one dimensional (in τ) evolution equation for tracer concentration along streamlines (Datta-Gupta and King, 1995). This equation is just as valid in one, two and three dimensions, and for homogeneous and heterogeneous media. The τ transformation includes all of these effects. All that is required for implementation is the velocity field and the calculation of the line integral (see Eq. (3)). The velocity field is typically obtained numerically using finite-difference method which makes the approach completely general. The streamline trajectories and the time of flight calculations are generally carried out using the Pollock algorithm (Pollock, 1988) and its extensions (Cordes and Kinzelbach, 1992; Jimenez et al., 2005).

2.2 Effective Density and Compressible Streamlines

For compressible multi-phase flow in porous media, the conserved quantity is a total multi-phase mass flux. Accordingly, we redefine the bi-streamfunctions to incorporate the compressibility effects (Bear, 1973; Cheng et al., 2006).

$$\rho \vec{u} = \nabla \psi \times \nabla \chi \tag{11}$$

where ρ represents an 'effective density' of the total fluids. For incompressible flow, we set $\rho = 1$ and we return to the standard incompressible streamline formulation. We can develop Eq. (11) further by recognizing that $\rho\vec{u}$ represents a conserved flux.

$$\begin{aligned} 0 = \nabla \bullet (\nabla\psi \times \nabla\chi) &= \nabla \bullet (\rho\vec{u}) \\ &= \vec{u} \bullet \nabla\rho + \rho\nabla \bullet \vec{u} \\ &= \phi\frac{\partial\rho}{\partial\tau} + \rho\nabla \bullet \vec{u} \end{aligned} \tag{12}$$

In Eq. (12) we have used the operator identity $\vec{u} \bullet \nabla = \phi\frac{\partial}{\partial\tau}$ to move from the physical space to the streamline time of flight coordinate τ. This coordinate transformation remains the same as that for incompressible flow as discussed later. The velocity field $\vec{u}$ is typically obtained from a finite difference solution and within each grid cell $\nabla \bullet \vec{u} = c$, a constant. Eq. (12) now reduces to an ordinary differential equation that can be easily integrated to obtain the variation of ρ along streamlines.

$$\rho = \rho_0 e^{-\left(c\frac{\tau}{\phi}\right)} \tag{13}$$

The value for the effective density can be traced along each streamline starting from the injectors where $\rho_0 = 1$ and where the initial volumetric flux $\Delta Q = \Delta\psi\,\Delta\chi$ is assigned to a streamline. This volumetric flux will vary along the streamline. It is now given by $\frac{1}{\rho}\Delta Q$ and the Darcy velocity is given by $\vec{u} = \frac{1}{\rho}\,\nabla\psi \times \nabla\chi$. Both depend upon the effective density. In the limit of small compressibility or small transit time, Eq. (13) has a simple approximation $\rho \approx \rho_0 - \rho_0\frac{\tau}{\phi}c$. This near linear variation with τ will be evident in the example given later.

For compressible flow, another modification needs to be made. This involves the Jacobian of the transformation from (x, y, z) to (τ, ψ, χ), to modify the relationship of volumes. This volumetric identity is used when mapping fluid volumes from streamlines to cells whenever streamlines need to be updated to account for changing conditions (Datta-Gupta, 2000). Correct treatment of this term is important to minimize mass balance errors. The generalization of Eq. (5) is now

$$\left\|\frac{\partial(\tau, \psi, \chi)}{\partial(x, y, z)}\right\| = (\nabla\psi \times \nabla\chi) \bullet \nabla\tau = \rho\vec{u} \bullet \nabla\tau = \rho\phi \tag{14}$$

Note that the operator identity in Eq. (8) still applies. However, in terms of volume we have,

$$\phi\; dx\; dy\; dz = \frac{1}{\rho} d\tau\; d\psi\; d\chi \tag{15}$$

The only remaining change in the formulation arises when developing the one dimensional transport equations along each streamline, where there is now an additional source/sink term in the 1-D equation to account for fluid expansion and compression along streamlines.

2.3 Compressible Streamline Formulation: An Illustration of the Method

We illustrate the basic steps in compressible streamline simulation using waterflood in a 2-D homogeneous 1/4 five-spot pattern under compressible flow conditions and compare the results with the current incompressible streamline formulation. The well configuration consists of a single water injection well and a single producing well and oil is the only mobile phase initially. We assume black-oil properties which imply that both water and oil are compressible and there is no interphase mass transfer between them. The water mass conservation equation for two-phase black oil is given by (Lake, 1989),

$$\phi \frac{\partial}{\partial t}\left(\frac{S_w}{B_w}\right) + \nabla \bullet \left(\frac{F_w \vec{u}_t}{B_w}\right) = 0 \tag{16}$$

In Eq. (16), B_w is the water formation volume factor accounting for water compressibility and F_w represents the fractional flow of water. Expanding the divergence operator we get,

$$\phi \frac{\partial}{\partial t}\left(\frac{S_w}{B_w}\right) + \frac{F_w}{B_w}\nabla \bullet \vec{u}_t + \vec{u}_t \bullet \nabla \left(\frac{F_w}{B_w}\right) = 0 \tag{17}$$

Now, transforming to streamline time of flight coordinates using, $\vec{u}_t \bullet \nabla = \phi \frac{\partial}{\partial \tau}$, and setting $\nabla \cdot \vec{u}_t = c$, we obtain

$$\frac{\partial}{\partial t}\left(\frac{S_w}{B_w}\right) + \frac{\partial}{\partial \tau}\left(\frac{F_w}{B_w}\right) = -\frac{c}{\phi}\frac{F_w}{B_w} \tag{18}$$

It is clear that compressibility effects generate source/sink terms along streamlines which account for fluid expansion/compression. Note that c is spatially varying along the streamline and can be obtained by mapping divergence of flux computed for each grid cell onto the streamline. For incompressible flow, B_w is constant and $c = 0$ everywhere and the right hand term vanishes resulting in the two-phase incompressible equation (Datta-Gupta and King, 1995).

Figure 1a shows the pressure distribution computed using finite-difference. The velocity field obtained from the pressure distribution is used to trace streamlines and compute the time of flight. Note that the Pollock algorithm (1988) is sufficiently general for this purpose and is not limited to incompressible flow. However, unlike incompressible flow, streamlines can now originate and terminate anywhere in the domain. The streamline time of flight for this 1/4 five-spot example is shown in Fig. 1b at t = 200 days. While tracing streamlines, we also compute the divergence of flux at each grid cell (Cheng et al., 2006) and the results are shown in Fig. 1c.

Next we calculate the effective densities along streamlines using Eq. (13). The tracing of the effective density along streamlines is shown in Fig. 2a. A contour of the 'local' changes in effective density ($\Delta\rho$ for each grid cell) is

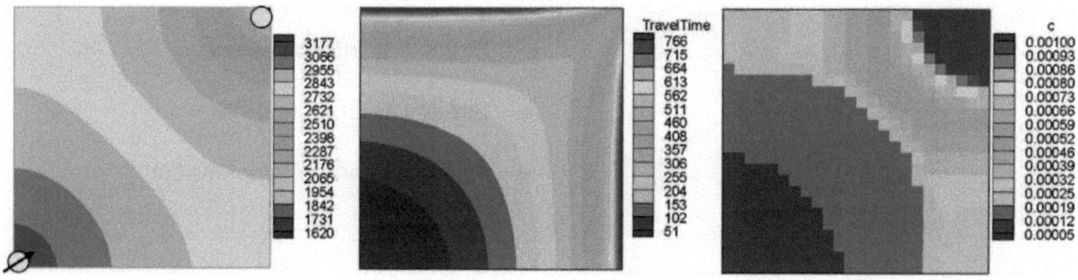

Fig. 1. Contour plots of pressure, time of flight and cell divergence of flux for waterflood in a 1/4 five-spot pattern

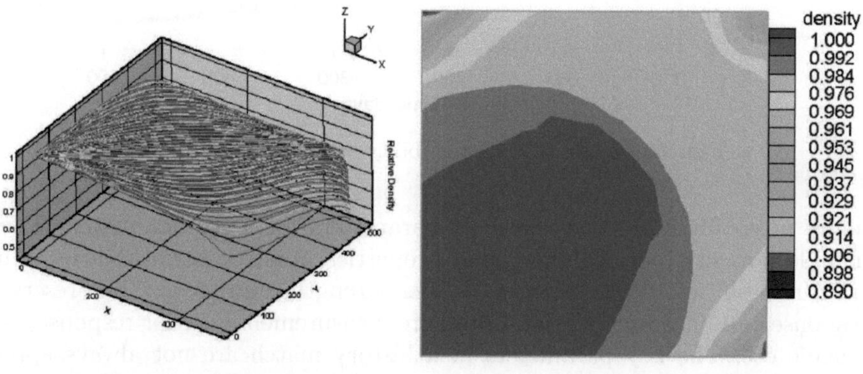

(a) Tracing effective density

(b) 'local' changes in effective density

Fig. 2. The tracing of effective density and contour showing changes

shown Fig. 2b. A value of less than unity indicates expansion of the fluid in the grid cell and vice versa. Note that the changes in effective density are a function of fluid compressibility, porosity and time of flight. The relatively low values at the stagnant corners reflect the large cell time of flight there. We can view the relative densities as scale factors for the time of flight, 'accelerating' or 'retarding' the particle transport along streamlines (see Eq. (15)).

The oil rate at the producing well for the compressible streamline calculations is shown in Fig. 3. For validation purposes, we have also shown the results from a commercial finite difference simulator. There is very good agreement between streamline and finite difference calculations. Finally, to demonstrate the effects of fluid compressibility, we have also superimposed the results from incompressible streamline formulation. Clearly, the compressibility effects are too large to be ignored for this case.

3 Role of Streamline Models in History Matching and Data Assimilation

In this section, we will discuss the role of streamline models in improving the workflow for history matching high resolution geologic models. There are

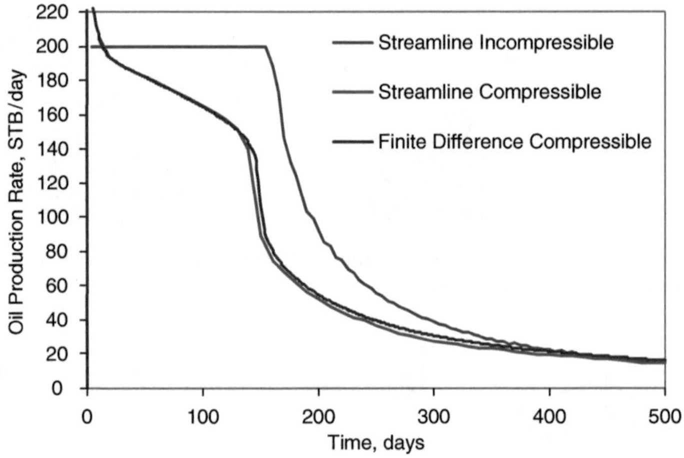

Fig. 3. Impact of compressibility on oil production rate

many possibilities for the choice of parameters for a history match. These include porosity, permeability, fluid properties, relative permeabilities, fault transmissibilities, fluid contacts, aquifer strength, and so on. The reservoir response can be water cut data, pressure measurements, tracer response, 4-D seismic etc. The key parameters in a history match are not always apparent. Also, the parameter and data uncertainties are often unknown and the constraints on the parameters are not well-defined. All these make field scale history matching a challenging and time consuming task.

Current industry practice in history matching generally follows a hierarchical workflow to account for uncertainties at various scales (Charles et al., 2001). To start with, typically a geologic model screening is carried out to identify the impact of large-scale features such as reservoir structure, fluid contacts, reservoir architecture/stratigraphy, and aquifer support on the production response. This step consists of performing flow simulations through a suite of realizations representing large-scale subsurface uncertainties. Best practice is to retain a sufficient number of conceptually distinct models to span these uncertainties. Streamline models are very useful at this model screening step because of their computational efficiency. The outcome of this step is a selected set of realizations that will undergo a more detailed history match. At this stage, the detailed history match will involve adjusting spatially variable properties such as permeability, porosity or facies distribution. This step involves localized changes and is typically the most time-consuming aspect of the workflow. Again, streamline models are quite advantageous at this stage. The spatial patterns defined by the streamlines can 'assist' in guiding these localized changes. In addition, streamline-based calculations may be performed to rapidly calculate sensitivities, which will allow an 'automatic' inversion. Streamline models can be used for both 'assisted' and 'automatic' history matching at this detailed stage as will be described below. At the last stage, we can examine the impact of physical properties or physical property model parameters around any specific history match (or matches). This stage typically involves adjusting physical properties such as the relative permeabilities and fluid properties. Because typically there are only a few parameters involved here, we can conveniently examine their significance using experimental design, which is a statistical technique that defines the most effective combinations of parameters to span the uncertainty space (Charles et al., 2001). The goal here is to minimize the number of simulations and at the same time to extract the maximum information within specified parameter ranges. Experimental design is typically followed by the creation of a 'response surface' and an 'analysis of variance' to examine the impact of model parameters on the production response. Again, streamline models can be helpful in the construction of the response surface because of their computational efficiency. The discussion that follows mainly focuses on the adjustment of spatially distributed properties, in particular permeability, for history matching using streamline models.

3.1 Streamline-based Assisted History Matching

Assisted history matching utilizes the streamline-based identification of injector-producer relationships to facilitate history matching (Emmanuel and Milliken, 1998; Agarwal and Blunt, 2003; Cheng et al., 2004). The method can be used with the actual simulation being performed by either finite difference or streamline simulator. The main steps in assisted history matching are: (i) simulation of production response using either a streamline or a finite-difference simulator (ii) tracing of streamlines and computation of the time of flight; (iii) use of streamlines to assign grid blocks or regions to each producer; (iv) computation of the mismatch between the observed and computed production response at each well; (v) updating the grid block or regional properties manually to improve the history match on a well-by-well basis. The use of streamlines leads to simple and unambiguous changes in the model. Also, because the changes are targeted and usually small, the geologic continuity tends to be preserved.

We now illustrate the procedure using a synthetic example. Figure 4a shows a 2D permeability field (50×50 grid) and the water cut response at four producing wells in a 5-spot pattern. We will treat this as our reference data. The initial or the prior model for permeability and the water cut responses are shown in Fig. 4b. The initial model was generated using the same variograms and histogram as the reference model. Thus, visually it appears quite close to the reference model, although the details are different. The flow responses, however, are quite different from the reference model.

Figure 5 shows the streamlines for the initial model. Streamlines are used to help assign grid cells to wells and to group the cells. From these streamlines, we know which grid cells to change to history match a particular well. This is illustrated for producer 4. In addition, from the time of flight we know which

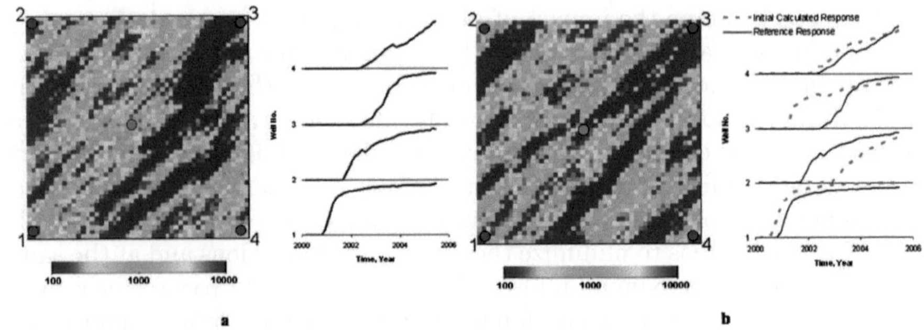

Fig. 4. (**a**) Reference permeability field and water cut response (**b**) Initial permeability field and water cut response

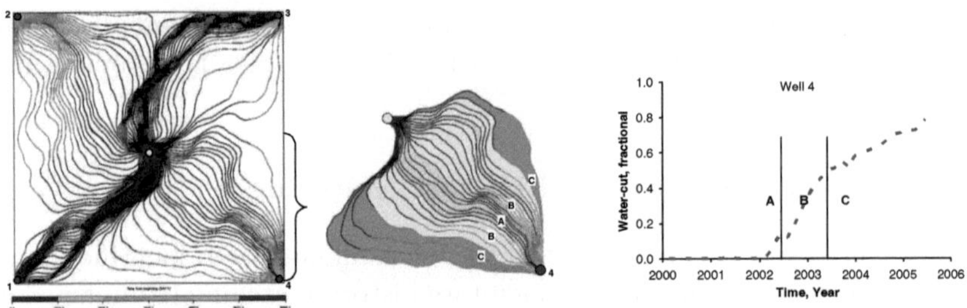

Fig. 5. An illustration of assisted history matching using streamlines

streamlines contribute to early stages (A), middle stages (B), and later stages (C) of the water cut. We can adjust permeability in cells covered by streamlines marked 'A' to match early breakthrough, and change those associated with 'B' and 'C' to match middle and later stages of water cut. Thus, assisted history matching (AHM) can accelerate the history matching process significantly. However, the method is still largely manual and requires considerable trial and error.

Difficulties in assisted history matching arise when there are changing field conditions such as infill drilling that will drastically alter the streamline patterns. The method then becomes less intuitive. Also, attempts to match some wells can cause matches in other wells to deteriorate because of the coupled nature of the flow field. Because, the permeability changes are made along streamlines, the method can introduce tube-like artifacts in the geologic model (Wang and Kovscek, 2000; Caers et al., 2002). Also, the approach will have difficulties in resolving features that are transverse to the dominant direction of the streamlines.

3.2 Streamline-based Inverse Modeling

Streamline-based automatic history matching or production data integration utilizes streamline-derived sensitivities to update geologic models based on production data (Vasco et al., 1999; Datta-Gupta et al., 2002; He et al., 2006). The sensitivities quantify the influence of reservoir properties on the production data. These sensitivities provide the fundamental relationships that allow us to invert the production data, measured at the wells, into modified reservoir properties between the wells.

The major steps are: (i) streamline-based flow simulation (ii) quantification of the production data mismatch; (iii) streamline-based analytic sensitivity computations and, (iv) updating of reservoir properties via inverse modeling.

To illustrate the procedure, we use the same synthetic example used for assisted history matching. The sensitivities relating the production data to changes in grid block permeability at the four producers are shown in Fig. 6. These sensitivities are calculated along the streamlines analytically during flow simulation and involve negligible computational cost. The details are discussed in the next section. Unlike assisted history matching, there is now no need to manually intervene to examine the streamline patterns to make changes to the model. The sensitivities can be used in conjunction with the iterative minimization algorithms to match the production data (Mclaughlin and Townley, 1996; Menke, 1989; Tarantola, 1987).

Figure 7 shows the updated permeability distribution, the matches to the water cut data and the permeability changes from the initial model to obtain the match. Clearly, the water cut responses are in good agreement.

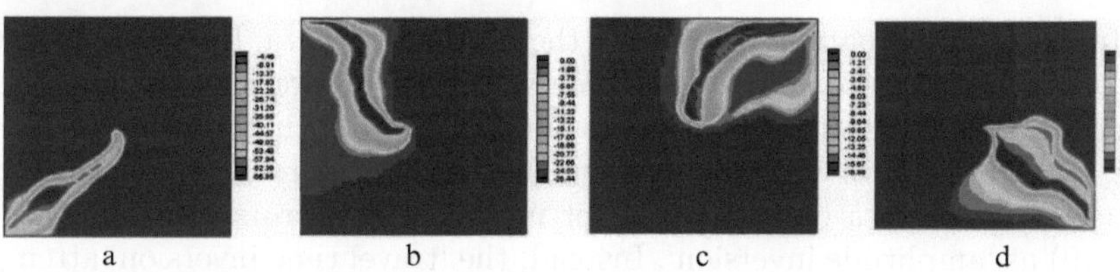

Fig. 6. Streamline-derived sensitivities for the four producing wells

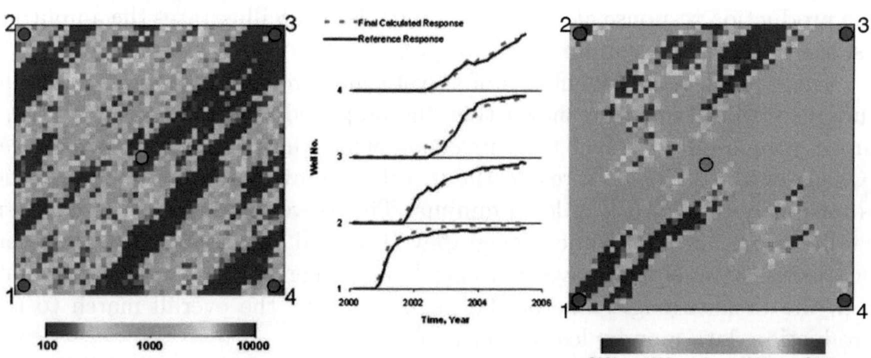

Fig. 7. Updated permeability field, water cut matches and changes in the permeability from the inversion

4 Streamline-based Inverse Modeling: Mathematical Background

In this section we discuss the mathematical details related to the quantification of data misfit and sensitivity computations in streamline-based inverse modeling. The sensitivity calculations include single and two-phase incompressible flow and three-phase compressible flow. We will demonstrate that the spatial patterns for the sensitivity analyses for the two-phase and the three-phase cases are closely related to that of single phase flow, and so we will cover it first. We will also cover in detail the travel time inversion of production data, and contrast it with more traditional 'amplitude' inversion.

4.1 Quantifying Data Misfit: Amplitude vs. Travel Time Inversion

Production data integration typically involves the minimization of a least squares functional representing the difference between the observed data and the calculated response from a simulator. The production data misfit is most commonly represented as follows

$$E_p = \sum_{j=1}^{N_w} \sum_{i=1}^{N_{dj}} w_{ij} \left(y_j^{cal}(t_i) - y_j^{obs}(t_i) \right)^2 \quad i = 1, \ldots, N_{dj} \quad j = 1, \ldots, N_w \quad (19)$$

In the above equation, y_j denotes the production data, for example water cut or tracer response at the producing well j, N_{dj} represents the number of observed data at well j, and N_w is the number of producing wells respectively. The w_{ij} are a set of weights that can be used to increase or decrease the influence of the data or to account for measurement errors. We will refer to Eq. (19) as 'amplitude inversion'. Instead, the 'travel time inversion' attempts to match the observed data and model predictions at some reference time, for example the breakthrough time or the peak arrival time. Thus, we will contrast the production response along the time axis. Figure 8 illustrates the amplitude vs. travel time inversion for tracer response in a well.

There are several advantages of travel time inversion compared to amplitude inversion. It can be shown that the amplitude inversion is highly nonlinear compared to travel time inversion which has quasi-linear properties (Cheng et al., 2005). As a result, the travel time inversion is more robust and is less likely to be stuck in local minima. This is well known in the geophysical literature and in seismic tomography (Lou and Schuster, 1991). However, because the travel time inversion entails matching only a single data point, e.g., the breakthrough time or the peak response, the overall match to the production data may be less than satisfactory.

We can actually combine the travel time inversion and amplitude inversion into one step via the 'generalized travel time inversion' (He et al., 2002). In this approach, we seek an optimal time-shift Δt of the data at each well so as

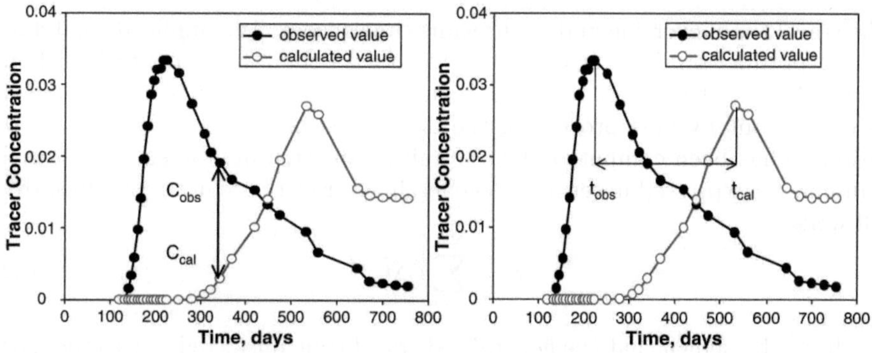

Fig. 8. An illustration of amplitude and peak travel time inversion

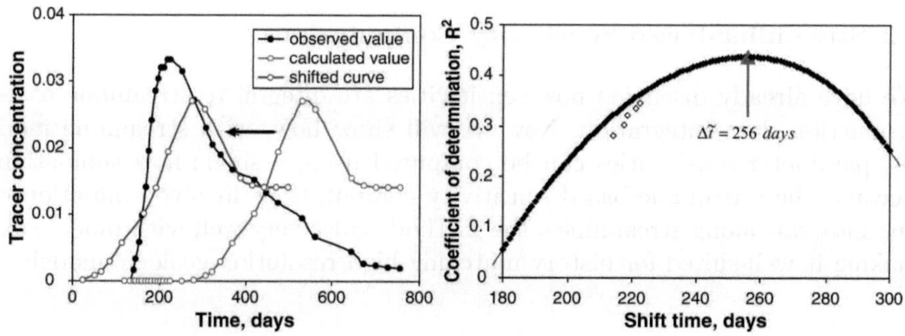

Fig. 9. An illustration of the calculation of generalized travel time

to minimize the production data misfit at the well. This is illustrated in Fig. 9 where the calculated tracer response is systematically shifted in small time increments towards the observed response and the data misfit is computed for each time increment. The optimal shift will be given by the Δt that minimizes the misfit function,

$$E\left(\Delta t\right)=\sum_{i=1}^{Nd}\left[y^{cal}\left(t_i+\Delta t\right)-y^{obs}\left(t_i\right)\right]^2 \tag{20}$$

Or, alternatively, we can maximize the coefficient of determination given by the following

$$R^2\left(\Delta t\right)=1-\frac{\sum\left[y^{cal}\left(t_i+\Delta t\right)-y^{obs}\left(t_i\right)\right]^2}{\sum\left[y^{obs}\left(t_i\right)-\overline{y^{obs}}\right]^2}. \tag{21}$$

Thus, the 'generalized travel time' at well j is given by the optimal time shift, $\Delta\tilde{t}_j$ that maximizes the correlation coefficient as illustrated in Fig. 9b. By defining the generalized travel time, we retain the desirable properties of the travel time inversion and at the same time accomplish amplitude matching of the production data. It is important to note that the computation of the optimal travel time shift does not require any additional flow simulation. It is carried out by post-processing the data at each well after the production response has been computed. The overall production data misfit for all wells can now be expressed in terms of a generalized travel time misfit summed over all wells.

$$E = \sum_{j=1}^{Nw} \left(\Delta \tilde{t}_j\right)^2 \tag{22}$$

It can be shown that the generalized travel time misfit reduces to the more traditional 'amplitude misfit' as we approach the solution (He et al., 2002).

4.2 Streamline-based Sensitivity Computations

We have already discussed how sensitivities are integral to streamline-based production data integration. Now we will show how, with streamline models, parameter sensitivities can be computed using a single flow simulation. Because the streamline-based sensitivity computations involve evaluation of line integrals along streamlines, the method scales very well with model size, making it well-suited for history matching high resolution geologic models.

Sensitivity to tracer travel time

We will first derive an expression for the sensitivity of the tracer travel time with respect to permeability and porosity for single phase incompressible flow. This tracer travel time is identical to the time of flight. The time of flight sensitivities can then be related to sensitivities of the tracer concentration or amplitude sensitivities. We start by rewriting the definition of the time of flight in terms of 'slowness',

$$\tau = \int_{\psi} s\left(\mathbf{x}\right) dr \tag{23}$$

where the integral is along the streamline trajectory Ψ, and $\mathbf{x}$ denote the position vector. The slowness s is defined as the reciprocal of the interstitial velocity $v(\mathbf{x})$, which is Darcy velocity divided by porosity,

$$s\left(\mathbf{x}\right) = \frac{1}{\left|v\left(\mathbf{x}\right)\right|} = \frac{\phi\left(\mathbf{x}\right)}{\left|u\left(\mathbf{x}\right)\right|} \tag{24}$$

Using Darcy's law, the slowness can now be written as,

$$s\left(\mathbf{x}\right) = \frac{\phi\left(\mathbf{x}\right)}{\lambda_{rt}\, k\left(\mathbf{x}\right) \left|\nabla P\right|} \tag{25}$$

Because s is a composite quantity involving reservoir properties, its first order variation will be given by,

$$\delta s(\mathbf{x}) = \frac{\partial s(\mathbf{x})}{\partial k(\mathbf{x})}\delta k(\mathbf{x}) + \frac{\partial s(\mathbf{x})}{\partial \phi(\mathbf{x})}\delta \phi(\mathbf{x}) \tag{26}$$

where the partial derivatives are

$$\frac{\partial s(\mathbf{x})}{\partial k(\mathbf{x})} \approx \frac{-\phi(\mathbf{x})}{\lambda_{rt}\,(k(\mathbf{x}))^2\,|\nabla P|} = -\frac{s(\mathbf{x})}{k(\mathbf{x})} \tag{27}$$

$$\frac{\partial s(\mathbf{x})}{\partial \phi(\mathbf{x})} = \frac{1}{\lambda_{rt} k(\mathbf{x})\,|\nabla P|} = \frac{s(\mathbf{x})}{\phi(\mathbf{x})} \tag{28}$$

In these variations, we are referring to changes in the magnitude of the permeability, i.e., changes for which the permeability anisotropy is preserved. The approximation in Eq. (27) and Eq. (28) is that the local perturbations in permeability generate negligible pressure changes (Eq. (28) is exact because steady state pressure distribution is independent of porosity). The implication of this assumption is that streamlines do not shift because of these small perturbations.

Now it is possible to calculate the change in travel time $\delta\tau$ to the change in slowness by integration along each streamline trajectory

$$\delta\tau = \int_{\psi} \delta s(\mathbf{x})\,dr = \int_{\psi}\left[-\frac{s(\mathbf{x})}{k(\mathbf{x})}\delta k(\mathbf{x}) + \frac{s(\mathbf{x})}{\phi(\mathbf{x})}\delta\phi(\mathbf{x})\right]dr \tag{29}$$

The tracer travel time sensitivity along a single streamline, Ψ with respect to permeability and porosity for a particular grid block at location $\mathbf{x}$ follows from Eq. (29) by simply carrying out the integral from the inlet to the outlet of the streamline within the grid block,

$$\frac{\partial\tau(\psi)}{\partial k(\mathbf{x})} = \int_{inlet}^{outlet}\left[-\frac{s(\mathbf{x})}{k(\mathbf{x})}\right]dr = -\frac{s(\mathbf{x})}{k(\mathbf{x})}\Delta r \tag{30}$$

$$\frac{\partial\tau(\psi)}{\partial \phi(\mathbf{x})} = \int_{inlet}^{outlet}\left[\frac{s(\mathbf{x})}{\phi(\mathbf{x})}\right]dr = \frac{s(\mathbf{x})}{\phi(\mathbf{x})}\Delta r \tag{31}$$

where Δr is the arc length of the streamline within the grid block. Notice that the sensitivity of the grid block permeability and porosity will be obtained by summing the contributions of all streamlines passing through the grid block.

For a conservative tracer with an injection concentration history given by $C_0(t)$, the response at the producer is given by the delay associated with the tracer travel time or time of flight along the streamline,

$$C(t) = C_0 \left(t - \int_{\psi} s(\mathbf{x}) dr \right) \tag{32}$$

The overall tracer response at the producer will be the sum of such contributions from all streamlines reaching the producer, diluted by the flux from those streamlines (Datta-Gupta and King, 1995). Using Eq. (32), we can derive an expression for sensitivity of tracer amplitude to porosity and permeability via chain rule and the travel time sensitivities (Vasco and Datta-Gupta, 1999).

Figures 10 and 11 compare the analytic sensitivities for tracer concentration in a 1/4 five-spot with the numerical perturbation method where each grid block was perturbed individually to compute the sensitivities. The numerical perturbation approach requires $M + 1$ simulations where $M = NX \cdot NY$ is the number of grid blocks compared to a single flow simulation for computing the analytic sensitivities. Overall, the agreement is very good. The slight differences for permeability sensitivities can be attributed to the stationary streamline assumptions in the analytic sensitivity computation. Because the pressure and streamline trajectories are not affected by porosity changes, the agreement between the analytic and numerical sensitivities is exact for this case.

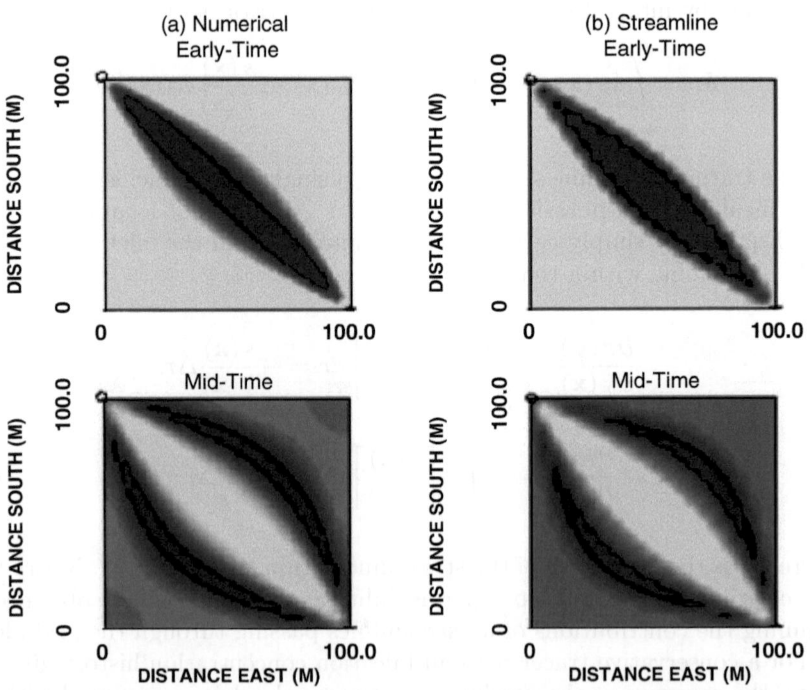

Fig. 10. Comparison of permeability sensitivities obtained using numerical perturbation and streamline models. Early time is at approximately 0.8 pore volumes injected, and mid-time is at approximately 1.2 pvi (Vasco and Datta-Gupta, 1999)

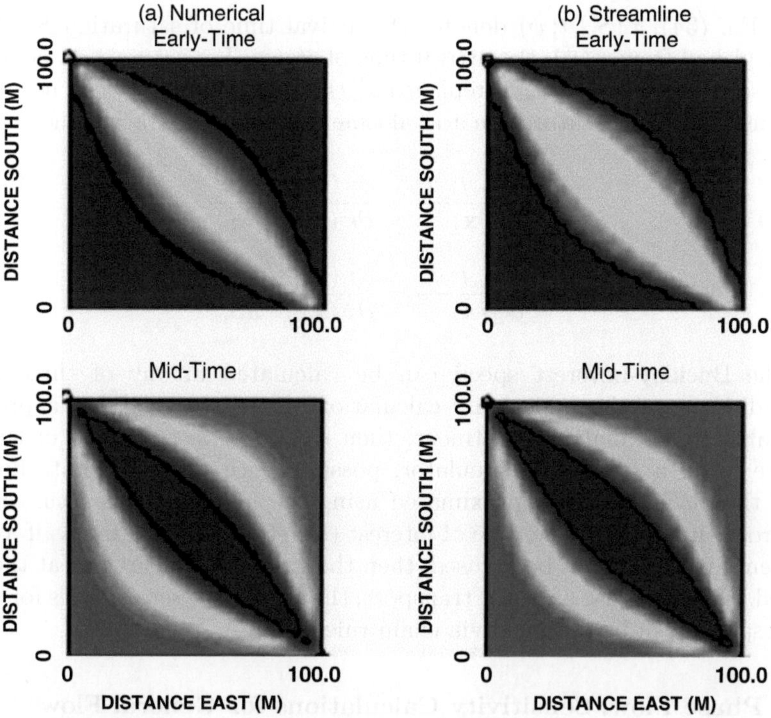

Fig. 11. Comparison of porosity sensitivities using numerical perturbation and streamline models. Early time is at approximately 0.8 pore volumes injected, and mid-time is at approximately 1.2 pvi (Vasco and Datta-Gupta, 1999)

The spatial pattern of the sensitivity depends upon the time at which it is calculated. Before breakthrough, the sensitivity in the domain is everywhere zero. After breakthrough, the non-zero sensitivity will spread as more streamlines breakthrough. At late times, the dominant sensitivity will be in the near-stagnant regions.

Two Phase Flow: Sensitivity of Saturation Front Arrival Times

Consider two-phase incompressible flow of oil and water described by the Buckley-Leverett equation in the streamline time of flight coordinates (Datta-Gupta and King, 1995),

$$\frac{\partial S_w}{\partial t} + \frac{\partial F_w}{\partial \tau} = 0 \tag{33}$$

The saturation velocity of a given saturation contour S_w along a streamline Ψ is given by the slope of the fractional flow curve,

$$\left(\frac{\partial \tau}{\partial t}\right)_{S_w} = \frac{dF_w}{dS_w} \quad or \quad \tau\left(S_w, t; \psi\right) = \tau\left(\psi\right) \Big/ \frac{dF_w}{dS_w} \tag{34}$$

In Eq. (34), $\tau(S_w, t; \psi)$ denotes the arrival time of saturation S_w, to be distinguished from $\tau(\psi)$, the arrival time of tracer. For saturations below the shock saturation, dF_w/dS_w is replaced by the shock speed. We can now relate the sensitivity of the saturation arrival time to that of the tracer time of flight.

$$\frac{\partial \tau(S_w, t; \psi)}{\partial k(\mathbf{x})} = \frac{\partial \tau(\psi)}{\partial k(\mathbf{x})} \bigg/ \frac{dF_w}{dS_w} \tag{35}$$

$$\frac{\partial \tau(S_w, t; \psi)}{\partial \phi(\mathbf{x})} = \frac{\partial \tau(\psi)}{\partial \phi(\mathbf{x})} \bigg/ \frac{dF_w}{dS_w} \tag{36}$$

The Buckley-Leverett speed can be calculated in any of three ways, depending upon the underlying calculation. If we are studying a problem amenable to an analytic treatment, then dF_w/dS_w is evaluated exactly. If we are using a streamline simulator, possibly with unsteady state flow effects, then dF_w/dS_w is approximated using the instantaneous saturation at the producing well, at the time of interest (He et al., 2002). Finally, if a finite difference calculation is being used, then the grid block saturation at the well is used. As in the case of tracer transport, the amplitude sensitivities for water cut response can be obtained via chain rule.

Two Phase Flow: Sensitivity Calculations for General Flow Geometry

The water cut sensitivity calculations become considerably more complicated for unsteady state flow involving rate changes, infill drilling etc when streamlines vary with time. One approach would be to resort to the numerical solution of the 1-D sensitivity equations along streamlines. These sensitivity equations can be derived by taking a direct derivative of the finite difference equations for the saturation evolution with respect to the variations in the reservoir parameters. There will be one such set of sensitivity equations to solve for each streamline for each time step (He et al., 2002; Gautier et al., 2001). In practice, the solution of these sensitivity equations can add significantly to the computational burden.

A more efficient alternative is obtained from an amplitude match via the generalized travel time inversion as discussed before. This requires computing the sensitivity of the generalized travel time for the water cut response as illustrated in Fig. 12.

Now, consider a small perturbation in reservoir properties, $\delta \mathbf{m}$ such that it results in a time-shift $\delta\tau$ for the entire computed water cut response at the producing well, that is, every data point has a common time-shift. We then have the following relationship for the observation times $(t_1, \ldots, t_{N_d})$

$$\delta\tau = \delta\tau_i = \left[\frac{\partial \tau_i}{\partial \mathbf{m}}\right]^T \delta\mathbf{m}, \quad i = 1, \ldots, N_d \tag{37}$$

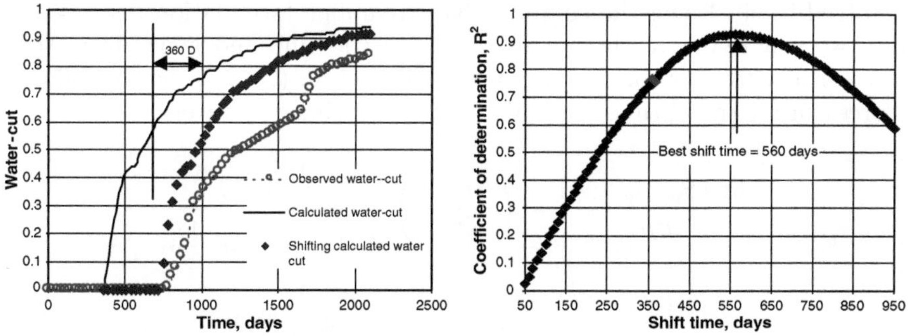

Fig. 12. Illustration of the generalized travel time calculation for the water cut

where **m** represents the reservoir parameter vector. Summing Eq. (37) over all the data points, we can arrive at the following simple expression for the sensitivity of the travel time shift with respect to the reservoir parameter (He et al., 2002),

$$\frac{\partial \tau}{\partial m} = \frac{1}{N_d} \sum_{i=1}^{N_d} \left(\frac{\partial \tau_i}{\partial m} \right) \tag{38}$$

Also, based on the definition of the generalized travel time, we have the following

$$\frac{\partial \Delta \tilde{t}}{\partial m} = -\frac{\partial \tau}{\partial m} \tag{39}$$

The negative sign in Eq. (39) reflects the sign convention adopted for defining the generalized travel time shift which is considered positive if the computed response is to the left of the observed data as shown in Fig. 12. We now obtain a rather simple expression for the sensitivity of the generalized travel time with respect to reservoir parameters as follows

$$\frac{\partial \Delta \tilde{t}}{\partial m} = \frac{1}{N_d} \sum_{i=1}^{N_d} \left(\frac{\partial \tau_i}{\partial m} \right) \tag{40}$$

In the above equation, the travel time sensitivity $(\partial \tau_i / \partial \boldsymbol{m})$ is given by Eq. (35) and Eq. (36). Notice that the travel time sensitivity calculation requires the derivative of the fractional flow, which is generally evaluated at the saturation of the outlet node for the streamline at the time of interest. This is an approximation that seems to work even under changing field conditions that require streamline updating (Cheng et al., 2004; He et al., 2002).

Three Phase Black Oil Case: Travel Time Sensitivities for Water cut and Gas-Oil Ratio

Three phase black oil properties assume that all three phases, water, oil and gas are compressible. In addition, gas can exist as a free phase or can be dissolved in oil. Using the compressible streamline formulation discussed before, we can re-derive the equations for travel time and generalized travel time sensitivities for three-phase flow. Cheng et al. (2006) gave the following expression for travel time sensitivities for matching the water cut,

$$\frac{\partial \tau (S_w, S_g, t; \psi)}{\partial m} = \left(\frac{\partial \tau (\psi)}{\partial m} \right) \frac{\frac{\partial}{\partial \tau}\left(\frac{S_w}{B_w}\right)}{\frac{\partial}{\partial \tau}\left(\frac{F_w}{B_w}\right) - \frac{F_w}{B_w}\frac{1}{\rho}\frac{\partial \rho}{\partial \tau}}$$

$$= \left(\frac{\partial \tau (\psi)}{\partial m} \right) \frac{\frac{\partial}{\partial \tau}\left(\frac{S_w}{B_w}\right)}{\frac{\partial}{\partial \tau}\left(\frac{F_w}{B_w}\right) + \frac{F_w}{B_w}\frac{c}{\phi}} \tag{41}$$

Similarly, the travel time sensitivity expression for gas-oil ratio (GOR) matching is given as follows (Cheng et al., 2006),

$$\frac{\partial \tau (S_w, S_g, t; \psi)}{\partial m} = \left(\frac{\partial \tau (\psi)}{\partial m} \right) \frac{\frac{\partial}{\partial \tau}\left(\frac{S_g}{B_g} + \frac{S_o R_s}{B_o}\right)}{\frac{\partial}{\partial \tau}\left(\frac{F_g}{B_g} + \frac{F_o R_s}{B_o}\right) + \left(\frac{F_g}{B_g} + \frac{F_o R_s}{B_o}\right)\frac{c}{\phi}} \tag{42}$$

In Eq. (42), S_0 and F_0 represent the saturation and fractional flow of oil. Similarly, S_g and F_g represent the saturation and fractional flow of the gas phase and R_s is the solution gas-oil ratio representing the amount of dissolved gas in the oil. Note that Eq. (41) reduces to the incompressible case (Eq. (35)) when ρ and B_w are constants. Also, the divergence free condition for incompressible flow requires, $c = 0$. As in the incompressible case, all the saturation dependent terms in Eqs. (41) and (42) are approximated using the instantaneous saturation at the producing well, at the time of interest. The spatial derivatives in τ are approximated using a backward difference.

Again, the sensitivities can be computed using a single flow simulation. Figure 13 compares the water cut sensitivities with respect to permeability for a 1/4 five-spot pattern using the streamline and the numerical perturbation methods for a black-oil case. The sensitivities for GOR are shown in Fig. 14.

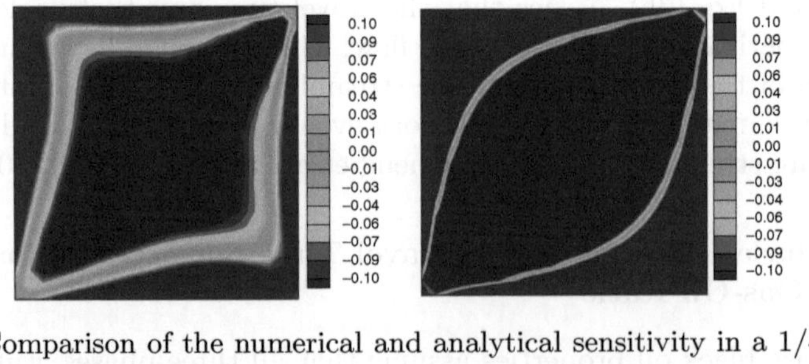

Fig. 13. Comparison of the numerical and analytical sensitivity in a 1/4-five spot pattern at a water cut of 0.5

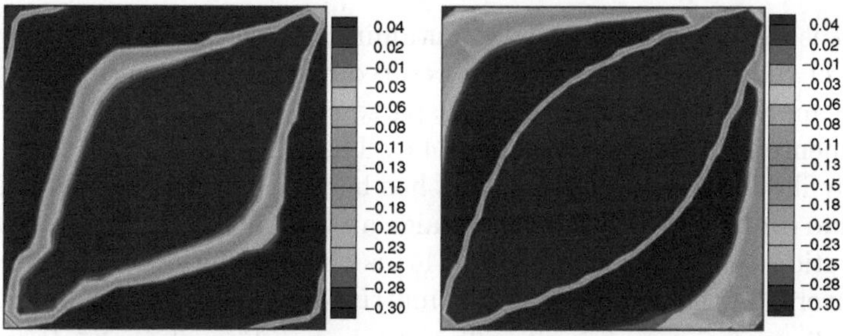

Fig. 14. Comparison of the numerical and analytical sensitivity in a 1/4-five spot pattern at a GOR of 4 Mscf/STB

It is obvious that the streamline-based sensitivities are only approximations because of the underlying assumptions of stationary trajectories. However, the trends are very similar. The analytic sensitivities are aligned along streamlines as expected.

It is worth amplifying upon the impact of the differences seen between the numerical and analytic calculations. The purpose of this calculation is to determine the sensitivities for an optimization procedure. The experience of Cheng et.al. (2006) is that the numerical sensitivities are sufficiently well approximated by the analytic calculations that the black oil history match converges. We will see this in the results section as well, in that a sufficient good approximation is all that is required for the history matching purposes.

4.3 Data Inversion

After the parameter sensitivities are computed, there are various approaches in the literature for the integration of production data via inverse modeling (Tarantola, 1987; Menke, 1989). These can be broadly classified into 'deterministic' and 'Bayesian' methods. Both methods have been successfully applied to history matching of field data. In this work, we have adopted a Bayesian formulation whereby we minimize the following penalized misfit function (Vega et al., 2004; Tarantola, 1987),

$$J(m) = \frac{1}{2}\left(m - m_p\right)^T C_M^{-1}\left(m - m_p\right) + \frac{1}{2}\left[\mathbf{\Delta}\tilde{\mathbf{t}}\right]^T C_D^{-1}\left[\mathbf{\Delta}\tilde{\mathbf{t}}\right] \tag{43}$$

In Eq. (43), $\mathbf{\Delta}\tilde{\mathbf{t}}$ is the vector of generalized travel time shift at the wells; C_D and C_M are the data error covariance and the prior model parameter covariance, respectively. The minimum in Eq. (43) can be obtained by an iterative least-squares solution to the linear system (Vega et al., 2004),

$$\begin{bmatrix} \mathbf{C}_D^{-1/2}\mathbf{G} \\ \mathbf{C}_\mathbf{M}^{-1/2} \end{bmatrix} \delta\mathbf{m} = \begin{bmatrix} \mathbf{C}_D^{-1/2}\left(\mathbf{\Delta}\tilde{\mathbf{t}}\right) \\ \mathbf{C}_\mathbf{M}^{-1/2}\left(\mathbf{m_p} - \mathbf{m}\right) \end{bmatrix} \tag{44}$$

where $\mathbf{G}$ is the sensitivity matrix containing the sensitivities of the generalized travel time with respect to the reservoir parameters and $\mathbf{m_p}$ represents the prior model. We use an iterative sparse matrix solver, LSQR, for solving the augmented linear system in Eq. (44). The LSQR algorithm is well suited for highly ill-conditioned systems and has been widely used for large-scale tomographic problems in seismology (Paige and Saunders, 1982). An important consideration in the solution of Eq. (44) is the calculation of the square-root of the inverse of the prior covariance matrix. We have used a numerical stencil that allows for an extremely efficient computation of $C_M^{-1/2}$ (Vega et al., 2004).

5 Application and Results

In this section, we demonstrate the application of the compressible streamline formulation and the streamline-derived sensitivities for inversion of multiphase production data, namely water cut and gas-oil ratio in three-phase flow conditions. Two examples are presented. The first one is a synthetic case to validate the approach. The second one is more involved and is included to demonstrate the feasibility of the approach for field-scale applications.

5.1 Inversion of Three-phase Flow Data: A Synthetic Example

This synthetic case involves reconstruction of a reference permeability field based on three-phase production history. We match water cut and GOR from a 9-spot pattern with the reference permeability distribution (Fig. 15).

The mesh size used is $21 \times 21 \times 1$. The reference permeability distribution consists of a low-permeability trend towards north and a high-permeability trend towards south. The water cut and GOR responses from the reference permeability field obtained from flow simulation are shown in Figs. 16 and 17. We treat these as the observed data.

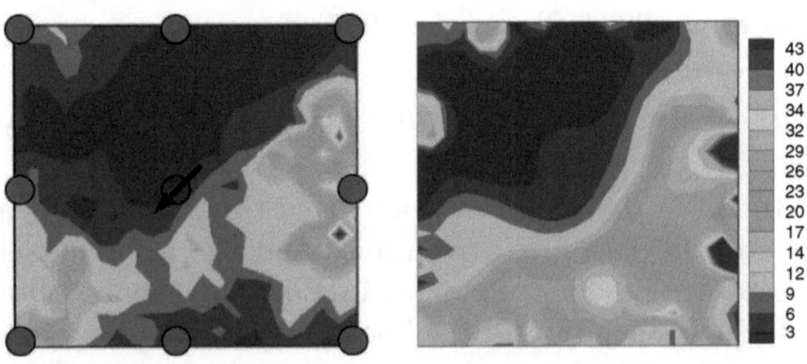

Fig. 15. History matching 3-phase flow for a 9-spot heterogeneous case: initial and final permeability field

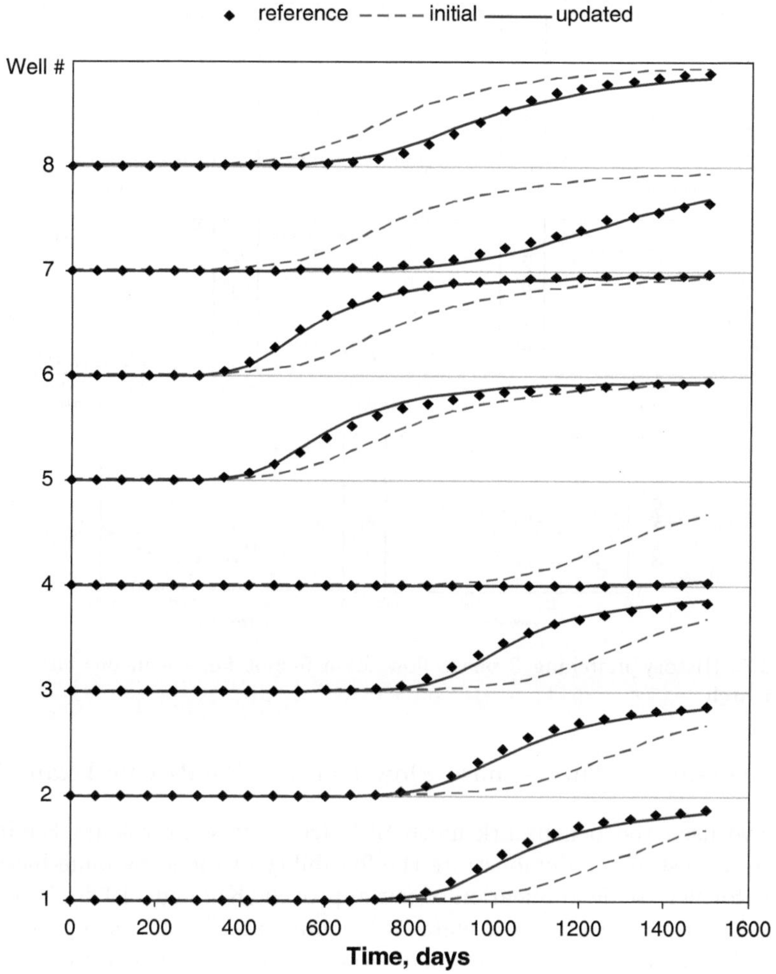

Fig. 16. History matching 3-phase flow for a nine-spot heterogeneous case: water cut match

Next, starting from a homogeneous initial permeability field we jointly match the water cut and GOR via our proposed generalized travel time inversion. The permeability for each grid block is treated as an adjustable parameter for this example (a total of 441 parameters). A comparison of the initial and final updated water cut matches is shown in Fig. 16, and that of GOR is in Fig. 17. Overall, the matches to the production data are quite satisfactory. The final permeability distribution is shown in Fig. 15. Clearly, the final permeability model captured the large-scale trend of the reference permeability field. The production data integration process here is very efficient and takes only a few iterations to converge.

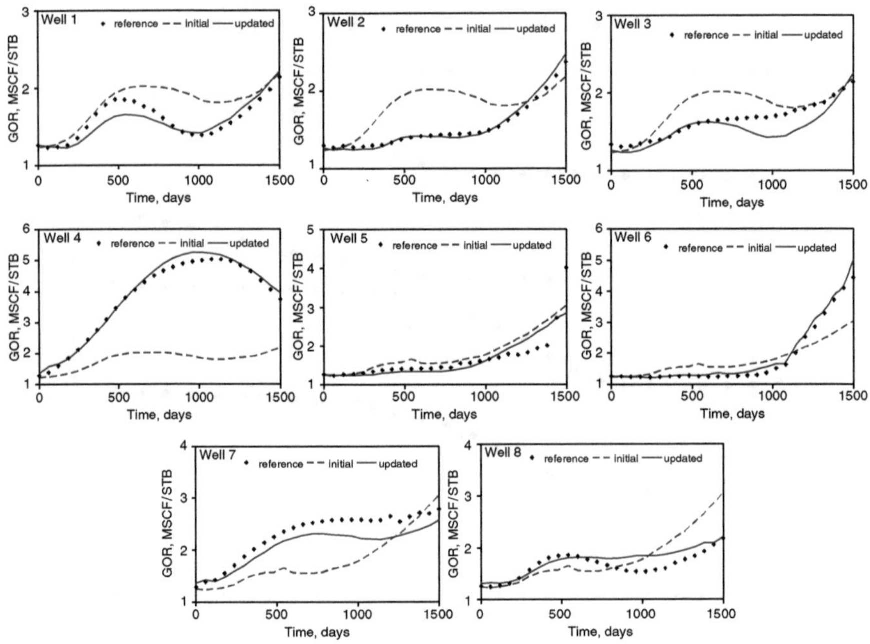

Fig. 17. History matching 3-phase flow for a 9-spot heterogeneous case: gas/oil ratio match

5.2 Inversion of Three-phase Flow Data: A Field-scale Example

We have used the benchmark ninth SPE (Society of Petroleum Engineers) comparative study to demonstrate the feasibility of the streamline-based approach for field-scale application of our approach (Killough, 1995). The ninth SPE comparative study investigates a bottom waterflood in a dipping reservoir with natural water encroachment from an aquifer. The reservoir (Fig. 18) is represented by a $24 \times 25 \times 15$ mesh with rectangular coordinates. The dimensions of the grid blocks are 300 feet in both the X- and Y-directions. Cell (1,1,1) is at a depth of 9000 feet subsea at the center of the cell top. The remaining cells dip in the X-direction at an angle of 10 degrees. Values of porosity and thickness can be found in the paper by Killough (1995). The total thickness from Layers 1 to 13 is 209 feet (16 feet per layer in average), and Layers 14 and 15 have thickness of 50 and 100 feet respectively. The oil/water contact is 9950 feet subsea. There is no free gas initially in the reservoir. After 900 days of production, there is considerable free gas saturation in the reservoir (Fig. 18).

A total of 26 wells, 1 water injector (I1) and 25 producers (named as P2 to P26) were included in the simulation model. The injector was completed from layers 1 through 11. All producers except producers 9, 17, 23, and 26 are completed in layers 1 to 13. Producers 9, 17, 23, and 26 are completed

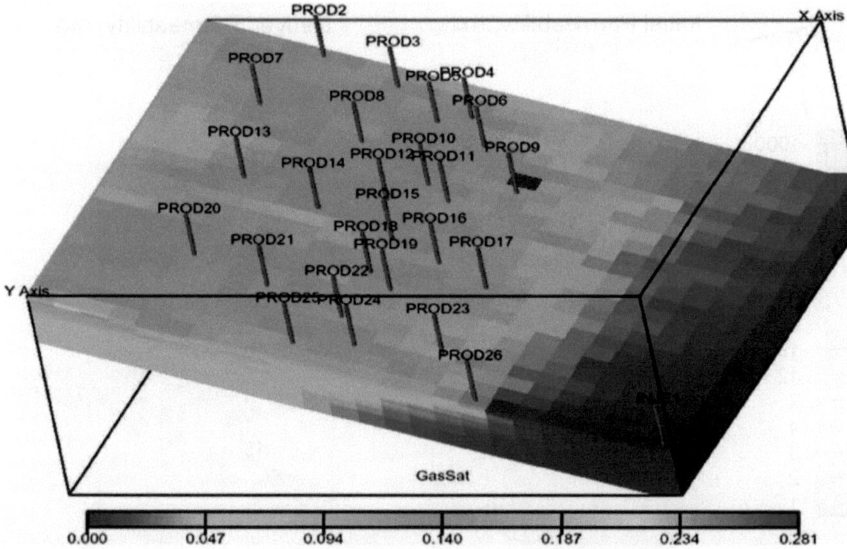

Fig. 18. Gas saturation distribution at the end of simulation time (900 days)

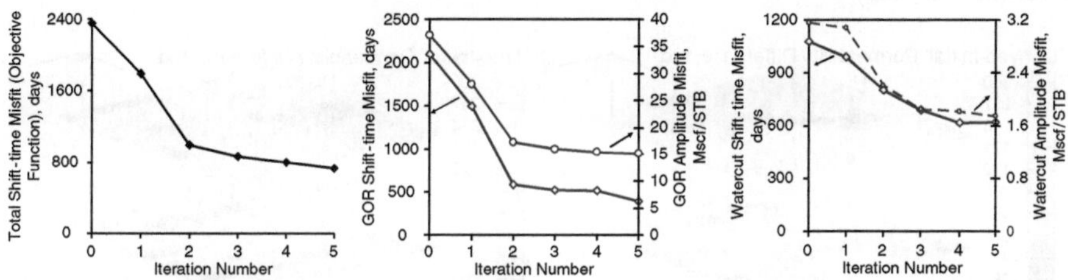

Fig. 19. Shift time and amplitude misfit reduction for joint water cut and GOR matching for SPE9 problem

in layers 1 to 5 so that no well will be perforated in the water leg. The water injector was injecting at a maximum bottomhole pressure of 4500 psia at a reference depth of 9110 feet subsea, and the producers were producing with a constant reservoir volume rate of 1800 RB/D and a minimum flowing bottomhole pressure of 1000 psia. The initial permeability field for inversion was generated via geostatistiscal simulation using data at the 26 well locations.

Figure 19 shows the convergence of the inversion algorithm. In 5 iterations, all misfit indices dropped appreciably. The misfit indices being the total misfit (GOR plus water cut shift time misfit), GOR shift time misfit, water cut shift-time misfit, GOR amplitude misfit, and water cut amplitude misfit. Most of the wells have a satisfactory match at the end of the inversion.

Figure 20 compares the initial permeability model and the updated (derived) model for the bottom five layers (Layers 11–15). The scale used is logarithmic and the minimum permeability is 0.003 md and the maximum is 10054 md. From a casual look, it is hard to discern the changes made to the ini-

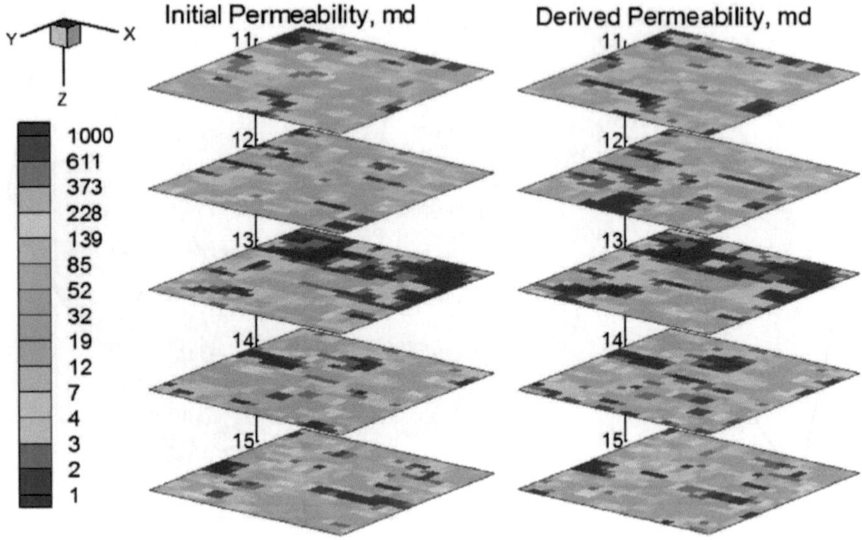

Fig. 20. Initial and derived (updated) permeability models for the SPE9 problem (bottom 5 layers)

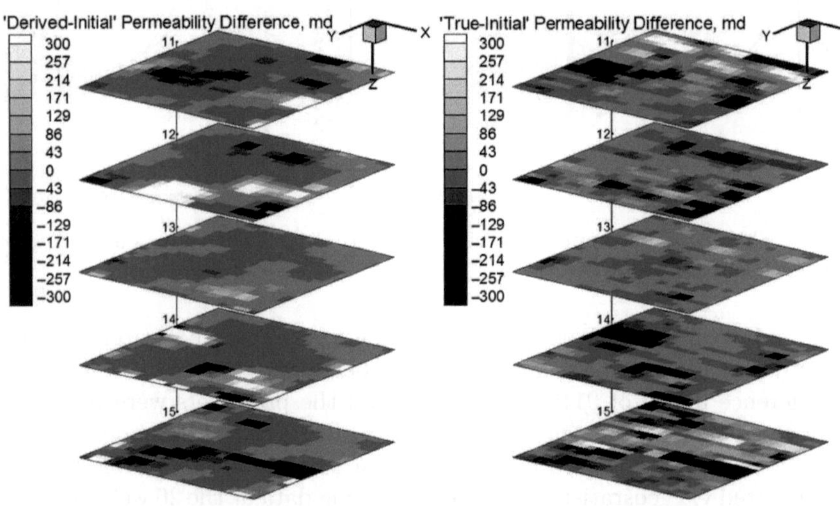

Fig. 21. Comparison of the "derived-initial" permeability difference and the "true-initial" permeability difference (bottom 5 layers)

tial model. This is because the inversion algorithm is designed to preserve the geologic continuity and the initial geologic features to the maximum possible extent. However, a careful comparison reveals many differences between the initial and the updated geologic models.

In Figure 21 the differences represent 'changes made'. This is to be compared with the 'changes needed' which is the difference between the reference

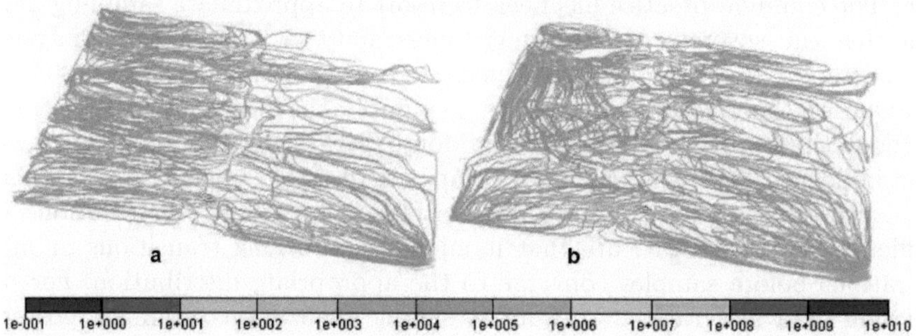

Fig. 22. (**a**) Streamline and time-of-flight distribution at the beginning of the simulation and (**b**) is the same at the end of the simulation for the SPE9 case

and the initial permeability model. We see that there is clearly very close agreement, particularly in regions where the permeability needs to be reduced (negative changes).

As might be expected, there are also some discrepancies. However, these are mostly in areas where the streamline density is low (Fig. 22) and there is not enough information to guide the changes (for example, near the boundary or in the areas where there is no well), or in the areas near the aquifer. For this field-scale case with 9000 parameters, it took about 50 minutes and 5 iterations to get a good history match using a PC with 1.5 GHz Pentium-4 processor.

6 Uncertainty Assessment Using Streamline Models

Traditional history matching usually generates a single matched model with unknown reliability, particularly for non-linear problems involving multiphase flow. Because of its computational efficiency, streamline-based history matching is well-suited for uncertainty assessment by generating multiple realizations of reservoir models that are conditioned to production data. The validity of the uncertainty quantification will strongly depend upon the distribution of these realizations, that is, whether they adequately represent the underlying uncertainties. In the context of the Bayesian inversion, the solution to the inverse problem is the posterior probability distribution itself. So, the problem of uncertainty quantification is closely tied to the appropriate sampling of multiple reservoir models from the posterior distribution. Such sampling is non-trivial because the posterior distribution has very high dimension (equal to the number of parameters) and is known only within a proportionality constant (un-normalized) (Tarantola, 1987). Furthermore, the posterior distribution can be both non-Gaussian and multi-modal. This makes rigorous sampling from the posterior distribution extremely computationally demanding. The common practice has been to resort to approximate sampling methods that can severely underestimate uncertainty. A comparative analysis of various approximate sampling schemes is given by Liu and Oliver (2003).

One of the rigorous approaches to sample the posterior distribution is through the use of the Markov Chain Monte Carlo (MCMC) algorithm (Omre and Lødøen, 2004). The MCMC methods are designed to generate a sequence of realizations that are samples from a target probability distribution. The difficulties with MCMC are that it might require long transitions or many iterations before samples converge to the appropriate distribution. For production data integration, we want to sample realizations from the posterior distribution, that is, $p_{m|d}(\cdot) \propto \exp(-J(m))$, where $J(m)$ is the misfit functional in Eq. (43). For each proposed realization of model parameters, for example, the permeability distribution, the computation of the acceptance probability requires flow simulation to compute $\boldsymbol{p}_{m|d}(\cdot)$. This can translate to thousands of simulations, particularly if the acceptance rate is low.

Efendiev et al. (2005) proposed a two-stage MCMC using an approximate likelihood calculation to improve the acceptance rate during MCMC sampling. Their approach does not compromise the rigor in traditional MCMC sampling. Instead, a pre-screening based on approximate likelihood calculations eliminates most of the rejected samples and the exact MCMC is performed only on the accepted proposals, with a higher chance of acceptance and without sacrificing the convergence characteristics. The approximate likelihood calculations should be fast and typically involves a linearized approximation around an already accepted state rather than an expensive computation such as a flow simulation.

Streamline models are particularly well-suited for this purpose because the analytic sensitivities provide a convenient means for computing the approximate likelihood function. For example, for a proposed transition $\boldsymbol{\delta}\mathbf{m} = \mathbf{m}^* - \mathbf{m}^\mathrm{i}$, we can relate the change in model parameters to the change in forward model response using the sensitivity matrix, $\boldsymbol{\delta}\mathbf{d} = \mathbf{G}\boldsymbol{\delta}\mathbf{m}$. Thus we can compute the first-stage acceptance probability based on the sensitivities computed from the previous accepted stage and without any flow simulation. A new simulation is performed only if the proposal is accepted at the first stage. This dramatically reduces the number of flow simulations required and makes production data integration and uncertainty assessment using MCMC feasible for field-scale applications.

Another recent development has been streamline-assisted Ensemble Kalman Filter (EnKF) for continuous reservoir model updating (Arroyo-Negrete et al., 2006). The EnKF is a Monte-Carlo approach that works with an ensemble of reservoir models (Naevdal et al., 2005). Specifically, the method utilizes cross-covariances between measurements and model parameters computed directly from the ensemble members to sequentially update the reservoir models. For practical field applications, the ensemble size needs to be kept small for computational efficiency. However, this leads to poor approximations of the cross-covariance matrix and loss of geologic realism through parameter overshoots, specifically by introducing localized patches of low and high permeabilities (Gu and Oliver, 2005). This difficulty is compounded by the strong non-linearity of the multiphase history matching problem.

Streamline trajectories and streamline-based sensitivities can be used in conjunction with the EnKF to modify the cross-covariance matrix to eliminate spurious covariance calculations arising from limited sample size. In particular, we can use flow-relevant information from streamlines to reduce the influence of distant observation points on model parameter updates (Arroyo-Negrete et al., 2006). It can be shown that the effect of such 'covariance localization' is to increase the effective ensemble size leading to an efficient and robust approach for history matching and continuous reservoir model updating. The streamline-assisted EnKF has been found to be quite general and avoids much of the problems in the traditional EnKF associated with instabilities, overshooting and loss of geologic continuity during model updating (Devegowda et al., 2007).

7 Summary and Conclusions

In this chapter we have highlighted the unique features of streamline models that make them particularly well-suited for production data integration into high resolution geologic models. Streamline models can be used for both 'assisted' and 'automatic' history matching and also in conjunction with finite-difference models. The unique information content in streamline trajectories, the time of flight and the streamline-derived sensitivities, allow for targeted changes in the geologic model to match production history. The changes are constrained to the prior model and thus geologic continuity is preserved.

We also examined the relative merits of 'travel time' vs. 'amplitude matching' for production data integration. In particular, the traditional amplitude inversion leads to a highly non-linear inverse problem with difficulties in stability and convergence. On the other hand, the travel time inversion is quasi-linear which makes it more robust and well-suited for field applications. One of the most important strengths of streamline models is their ability to compute parameter sensitivities analytically using a single flow simulation. This is the single-most important feature that makes streamline models extremely well-suited for history matching high resolution geologic models via inverse modeling. The sensitivities are simple integrals along streamlines and can be easily calculated for single phase. These results generalize easily to two phase incompressible flow and also for the three-phase black oil case.

We have briefly discussed uncertainty quantification during history matching using Markov Chain Monte Carlo (MCMC) methods. In particular, we have seen that the streamline models allow for an efficient multistage MCMC with higher acceptance rates while preserving the rigor of sampling from the posterior distribution. Finally, we introduce the role of streamlines for flow-relevant covariance localization during sequential data assimilation using the Ensemble Kalman Filter.

References

Agarwal A, Blunt MJ (2003) Streamline-Based Method with Full-Physics Forward Simulation for History-Matching Performance Data of a North Sea Field. SPE Journal 8(2):171–180

Al-Harbi M et al. (2005) Streamline-based Production Data Integration in Naturally Fractured Reservoirs. SPE Journal 10(4): 426–439

Anderman ER, Hill MC (1999) A new multistage groundwater transport inverse method: Presentation, evaluation, and implications. Water Resour Res 35(4): 1053–1063

Arroyo-Negrete E, Devegowda D and Datta-Gupta A (2006) Streamline Assisted Ensemble Kalman Filter for Rapid and Continuous Reservoir Model Updating. SPE 104255 in Proceedings of the SPE International Oil and Gas Conference and Exhibition, Beijing, China

Bear J (1973) Dynamics of Fluid in Porous Media, Dover Publications, New York

Bratvedt F, Gimse T, Tegnander C (1996) Streamline Computations for Porous Media Flow Including Gravity. Transport Porous Med 25:63

Caers J et al (2002) A Geostatistical Approach to Streamline Based History Matching. SPE Journal 7(3):250–266

Charles T et al (2001) Experience with the Quantification of Subsurface Uncertainties. SPE 68703 in Proceedings of the Asia Pacific Oil and Gas Conference and Exhibition, Jakarta, Indonesia

Cheng H et al (2006) Compressible Streamlines and Three-phase History Matching. SPE 99465 in Proceedings of the SPE/DOE Symposium on Improved Oil Recovery, Tulsa, Oklahoma

Cheng H et al (2006) A Rigorous Compressible Streamline Formulation for Two and Three Phase Black Oil Simulation. SPE Journal 11(4): 407–417

Cheng H, Datta-Gupta A, He Z A (2005) Comparison of Travel Time and Amplitude Inversion for Production Data Integration into Geologic Models: Sensitivity, Nonlinearity and Practical Implications. SPE Journal 10(1):75–90

Cheng H et al (2004) Field Experiences with Assisted and Automatic History Matching Using Streamline Models. SPE 89857 in Proceedings of the SPE Annual Technical Conference and Exhibition, Houston, Texas

Cordes C, Kinzelbach W (1992) Continuous Groundwater Velocity Fields and Pathlines in Linear, Bilinear and Trilinear Finite Elements. Water Resour Res 28(11):2903

Datta-Gupta A, King MJ (1995) A Semianalytic Approach to Tracer Flow Modeling in Heterogeneous Permeable Media. Adv Water Resour 18(1): 9

Datta-Gupta A (2000) Streamline Simulation: A Technology Update. SPE Distinguished Author Series Journal of Pet. Tech. 68–73

Datta-Gupta A et al (2001) Streamlines, Ray Tracing and Production Tomography: Generalization to Compressible Flow. Pet Geosci 7:S75–S86

Datta-Gupta A, Yoon SS, Pope GA., Vasco DW (2002) Inverse Modeling of Partitioning Interwell Tracer Tests: A Streamline Approach. Water Resour Res 30 (6):10.1029

Devegowda D, Arroyo-Negrete E, Datta-Gupta A, Douma SG (2007) Efficient and Robust Reservoir Model Updating Using Ensemble Kalman Filter with Sensitivity-Based Covariance Localization. SPE 106144 in Proceedings of the SPE Reservoir Simulation Symposium, Houston, Texas

Dubrule O (1998) Geostatistics in Petroleum Geology. American Association of Petroleum Geology, Course note series 38

Efendiev Y et al (2005) An Efficient Two-Stage Markov Chain Monte Carlo Method for Dynamic Data Integration. Water Resour Res 41:W12423, 10.1029/2004WR003764

Emanuel AS, Milliken WJ (1998) History Matching Finite-Difference Models with 3-D Streamlines. SPE 49000 in Proceedings of the SPE Annual Technical Conference and Exhibition, New Orleans, Louisiana

Gautier Y, Nœtinger B, Roggero F (2001) History Matching Using a Streamline-Based Approach and Exact Sensitivity Coefficients. SPE 71626 in Proceedings of the SPE Annual Technical Conference and Exhibition, New Orleans, Louisiana

Gu Y and Oliver DS (2005) History Matching of the PUNQ-S3 Reservoir Model Using the Ensemble Kalman Filter. SPE Journal 10(2):217–224

He Z, Yoon S, Datta-Gupta (2002) A Streamline-Based Production Data Integration Under Changing Field Conditions. SPE Journal 7(4): 423–436

He Z, Datta-Gupta,A., Vasco DW (2006) Rapid Inverse Modeling of Pressure Interference Tests Using Trajectory-based Travel Time and Amplitude Matching. Water Resour Res 42:W03419 doi:10.1029/2004WR003783

Hyndman DW, Harris JM, Gorelick SM (1994) Coupled seismic and tracer test inversion for aquifer property characterization. Water Resour Res 30(7):1965–1977

Jimenez E et al (2005) Spatial Error and Convergence in Streamline Simulation. In: Proceedings of the SPE Reservoir Simulation Symposium, The Woodlands, Texas

Killough JE (1995) Ninth SPE Comparative Solution Project: A Reexamination of Black-Oil Simulation. In: Proceedings of the SPE Reservoir Simulation Symposium, San Antonio, Texas

King MJ, Datta-Gupta A (1998) Streamline Simulation: A Current Perspective. In Situ 22:1

Kitanidis PK (1995) Quasilinear Geostatistical Theory for Inversing. Water Resources Research 31(10):2411–2419

Lake LW (1989) Enhanced Oil Recovery. Prentice Hall, Englewood Cliffs, New Jersey

Liu N, Oliver DS (2003) Evaluation of Monte Carlo Methods for Assessing Uncertainty. SPEJ 8(2):188–195

Lou Y, Schuster GT (1991) Wave Equation Traveltime Inversion. Geophysics 56(5): 645–653

Ma X, Al-Harbi M, Datta-Gupta A, Efendiev Y (2006) A Multistage Sampling Approach to Quantifying Uncertainty During History Matching Geological Models, SPE 102476 presented at the SPE Annual Technical Conference and Exhibition held in San Antonio, Texas, U.S.A.

Mclaughlin D, Townley LR (1996) A Reassessment of the Groundwater Inverse Problem. Water Resour Res 32(5): 1131–1161

Medina A, Carrera J (1996) Coupled estimation of flow and solute transport parameters. Water Resources Research 32(10):3063–3076

Menke W (1989) Geophysical Data Analysis: Discrete Inverse Theory. Academic Press Inc, California

Naevdal et al. (2005) Reservoir Monitoring and Continuous Model Updating Using Ensemble Kalman Filter. SPE Journal 10(1):66–74

Oliver DS et al (2001) Integration of Production Data Into Reservoir Models. Pet Geosci 7:S65–S73

Omre H, Lødøen OP (2004) Improved Production Forecasts and History Matching Using Approximate Fluid-Flow Simulators SPEJ 9(3):339–351

Osako I, Datta-Gupta A (2007) A Compositional Streamline Formulation with Compressibility Effects. In: Proceedings of the SPE Reservoir Simulation Symposium, Houston, Texas

Paige CC and Saunders MA (1982) LSQR: An Algorithm for Sparse Linear Equations and Sparse Least Squares ACM Trans Math Soft 8(1):43–71

Pollock DW (1988) Semianalytical Computation of Pathlines for Finite-Difference Models. Groundwater 26(6):743

Tarantola A (1987) Inverse Problem Theory. Elsevier, Amsterdam, The Netherlands

Vasco DW, Yoon S, Datta-Gupta A (1999) Integrating Dynamic Data into High Resolution Reservoir Models Using Streamline-Based Analytic Sensitivity Coefficients. SPEJ 4(4): 389–399

Vasco DW, Datta-Gupta A (1999) Asymptotic Solutions for Solute Transport: A Formalism for Tracer Tomography. Water Resour Res 35:1–16

Vega L, Rojas D, Datta-Gupta A (2004) Scalability of the Deterministic and Bayesian Approaches to Production Data Integration Into Field-Scale Reservoir Models. SPE Journal 9(3):330–338

Wang Y and Kovscek AR (2000) Streamline Approach for History Matching Production Data. SPE Journal 5(4):353–362

Yeh T, Jim C, Liu S (2000) Hydraulic tomography: development of a new aquifer test method. Water resources Research 36(8):2095–2105

Information Fusion in Regularized Inversion of Tomographic Pumping Tests

Geoffrey C. Bohling

Kansas Geological Survey, University of Kansas

Summary. In this chapter we investigate a simple approach to incorporating geophysical information into the analysis of tomographic pumping tests for characterization of the hydraulic conductivity (K) field in an aquifer. A number of authors have suggested a tomographic approach to the analysis of hydraulic tests in aquifers – essentially simultaneous analysis of multiple tests or stresses on the flow system – in order to improve the resolution of the estimated parameter fields. However, even with a large amount of hydraulic data in hand, the inverse problem is still plagued by non-uniqueness and ill-conditioning and the parameter space for the inversion needs to be constrained in some sensible fashion in order to obtain plausible estimates of aquifer properties. For seismic and radar tomography problems, the parameter space is often constrained through the application of regularization terms that impose penalties on deviations of the estimated parameters from a prior or background model, with the tradeoff between data fit and model norm explored through systematic analysis of results for different levels of weighting on the regularization terms. In this study we apply systematic regularized inversion to analysis of tomographic pumping tests in an alluvial aquifer, taking advantage of the steady-shape flow regime exhibited in these tests to expedite the inversion process. In addition, we explore the possibility of incorporating geophysical information into the inversion through a regularization term relating the estimated K distribution to ground penetrating radar velocity and attenuation distributions through a smoothing spline model.

1 Introduction

A number of investigators (Neuman, 1987; Tosaka et al., 1993; Bohling, 1993; Gottlieb and Dietrich, 1995; Butler et al., 1999; Yeh and Liu, 2000; Vesselinov et al., 2001a, 2001b; Bohling et al., 2002; Liu et al., 2002; Brauchler et al., 2003; Zhu and Yeh, 2005) have proposed hydraulic tomography as a means for obtaining higher-resolution estimates of distributions of aquifer flow and transport properties than can be obtained from traditional aquifer tests. Hydraulic tomography involves performing a sequence of pumping or slug

G.C. Bohling: *Information Fusion in Regularized Inversion of Tomographic Pumping Tests*, Studies in Computational Intelligence (SCI) **79**, 137–162 (2008)
www.springerlink.com

tests stressing different vertical intervals in an aquifer and measuring pressure or head responses at a number of different observation locations. This testing sequence yields a dataset that has "sampled" the aquifer over a number of different streamlines or flowpaths, similar to the crossing raypath pattern used in seismic or radar tomography (Aster et al., 2005). Simultaneous inversion of the entire dataset allows the development of an estimated distribution of hydraulic properties at a higher resolution than can be obtained from traditional approaches such as large-scale pumping tests (Butler, 2005).

Even with a large amount of data from multiple tests in hand, the inverse problem is still plagued by the non-uniqueness (numerous property distributions produce essentially the same observed response) and ill-conditioning (small changes in the data produce large changes in estimated parameters) that plague many parameter estimation problems in the earth sciences (Carrera and Neuman, 1986; Parker, 1994; Aster et al., 2005). As a result, the parameter space for the inversion needs to be restricted in a sensible fashion in order to obtain plausible parameter estimates. A common approach to restricting the parameter space in groundwater flow and transport problems is zonation, where the properties are restricted to constant values within prespecified zones, based on a geological model and auxiliary information such as drillers' logs or geophysical well logs. Another approach is to condition the inversion on a prior model developed through geostatistical estimation or simulation techniques. Carrera et al. (2005) provide a recent review of various approaches to incorporating auxiliary information into the hydrogeological inverse problem.

A number of investigators have proposed the incorporation of geophysical information in the groundwater inverse problem, as summarized in the recent volume edited by Rubin and Hubbard (2005). In these studies, estimated distributions of geophysical properties derived from geophysical surveys are used to infer information regarding the distribution of hydrogeological parameters, either by providing information about the spatial structure of those parameters (in other words, to help develop an appropriate zonation for the hydrogeological inverse problem) or through the exploitation of presumed functional relations or correlations between the geophysical and hydrogeological parameters. Due to the lack of universally valid deterministic relationships between geophysical and hydraulic parameters, the incorporation of geophysical information is necessarily a site-specific endeavor, usually based on statistical relationships between geophysical parameters and limited measurements of hydraulic parameters (Hyndman and Tronicke, 2005).

Bohling et al. (2007) present an analysis of a set of tomographic pumping tests in an alluvial aquifer based on layered zonations of the aquifer hydraulic conductivity (K) derived from crosshole ground penetrating radar surveys. These zonations were based on a cluster analysis of radar velocity and attenuation profiles, similar to the approach presented by Tronicke et al. (2004). In this study, we examine the same data as Bohling et al. (2007) using a different approach, attempting to exploit possible quantitative correlations between K and the radar velocity and attenuation. Rather than specifying the nature of the functional relationship in advance, we incorporate the geophysical data through a regularization term in the objective function for the hydraulic inversion. This regularization term penalizes deviations between the current K estimates and a best-fit adaptive smoothing spline model relating those estimates to the radar velocity and attenuation. In other words, the term measures how well the estimated K distribution can be represented as a smooth function of the geophysical parameters, without requiring an *a priori* specification of the exact nature of the functional relationship. Increasing the weight on this regularization term will force the K distribution to more closely match the distributions of the geophysical parameters. Thus, this approach requires an *a posteriori* assessment of the plausibility of the resulting relationship between K and the geophysical attributes.

To serve as a background for the regularization relative to geophysical parameters, we first examine regularized inversion relative to a uniform background model, applying zeroth-order Tikhonov regularization, a technique commonly employed in geophysical tomography (Aster et al., 2005), to the hydraulic tomography problem. This follows the suggestion of Tonkin and Doherty (2005) and Doherty (2003), who encourage the use of Tikhonov regularization in hydrogeological inverse problems. This kind of regularization results in a constrained least squares inversion, a method that has been employed for decades in groundwater inverse problems. However, in keeping with a more geophysical approach to the problem we here emphasize a systematic exploration of the tradeoff between data fit and deviations from the prior model. We then extend the approach with a regularization term incorporating the geophysical information.

2 Experimental Setup

Here we will briefly summarize the tomogrpahic pumping tests examined in this study. A more detailed description is provided in Bohling et al. (2007) and Butler (2005) provides an overview of various kinds of hydraulic tests performed at the test site. The tests were performed at the Kansas Geological Survey's Geohydrologic Experimental and Monitoring Site (GEMS), in the Kansas River valley in northeast Kansas, USA. The alluvial aquifer at the site consists of approximately 10.5–11 meters of highly transmissive sand and gravel overlain by approximately 11 meters of silt and clay. The silt and clay serve to confine the alluvial aquifer over the time periods involved in the hydraulic tests at the site. The overall average hydraulic conductivity of the site, derived from large-scale pumping tests and from Cooper-Jacob (Cooper and Jacob, 1946) analyses of the data described here (Bohling et al., 2007) is approximately 130 m/day or 0.15 cm/s, a value in the expected range for clean sand (Freeze and Cherry, 1979).

We performed a series of 23 tomographic pumping tests by isolating and pumping intervals in two wells, each 11 centimeters in diameter. The two wells, Gems4N and Gems4S, are separated by about 10 meters, as shown in Fig. 1, with Gems4N on the north end and Gems4S on the south end of the vertical cross-section of aquifer that is the focus of this investigation. We performed 11 tests with pumping in Gems4N and 12 with pumping in Gems4S, each involving pumping over a 0.6-meter interval. The sequence of tests, numbered from the bottom up in each well, is represented by the line segments on the left and right sides of Fig. 1. During each pumping test, drawdowns were measured using pressure transducers installed in two multichamber PVC samplers located between Gems4N and Gems4S. Each multichamber sampler consists of a central chamber and six surrounding chambers. Each of these six chambers contains a sampling port at a different depth, giving a total of 12 sampling ports over the vertical section, whose locations are also shown in Fig. 1. The six pressure transducers employed for these tests were located either in the even- or odd-numbered sample ports, numbered from the bottom up, during each pumping test. The tests using odd-numbered sample ports

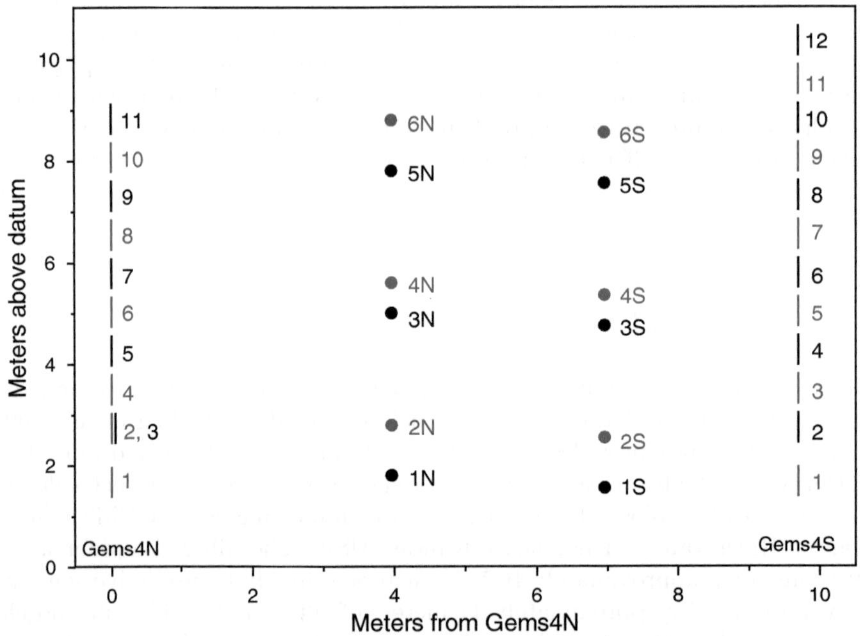

Fig. 1. GEMS tomographic test setup. Lines on left and right indicate sequence of pumping intervals in two different pumping wells, Gems4N and Gems4S, and circles indicate drawdown observation points. Tests indicated with black lines used six observation points marked in black and tests marked in gray used observation points marked in gray.

are represented in black in Fig. 1 and those using even-numbered ports are represented in gray.

Each test involved 900 s (15 min) of pumping followed by a recovery period of equal duration and an additional 15 or more minutes of preparation for the next test. This allowed ample time for drawdowns to return to static (pre-test) levels between tests.

During each test, drawdowns were measured at a rate of 2 samples per second for the duration of pumping and recovery. Figure 2 shows a subset of the data, measurements at two-second intervals over the interval from 20 seconds to 70 seconds after initiation of pumping, for the 12 pumping tests in Gems4S. Each panel in Fig. 2 shows the data obtained at the six sampling ports in each test, with the tests numbered from the bottom up as in Fig. 1. Each line of points represents the drawdown from a different sampler, with the largest drawdowns in each test corresponding to samplers closest to the pumping interval. The data from the 11 tests with pumping in Gems4N are similar in character. An obvious feature of these data is that they display a roughly constant and common slope and roughly constant differences between drawdowns at different samplers in each test. This is the signature of a steady-shape flow regime (Kruseman and de Ridder, 1990; Bohling et al., 2002). Apart

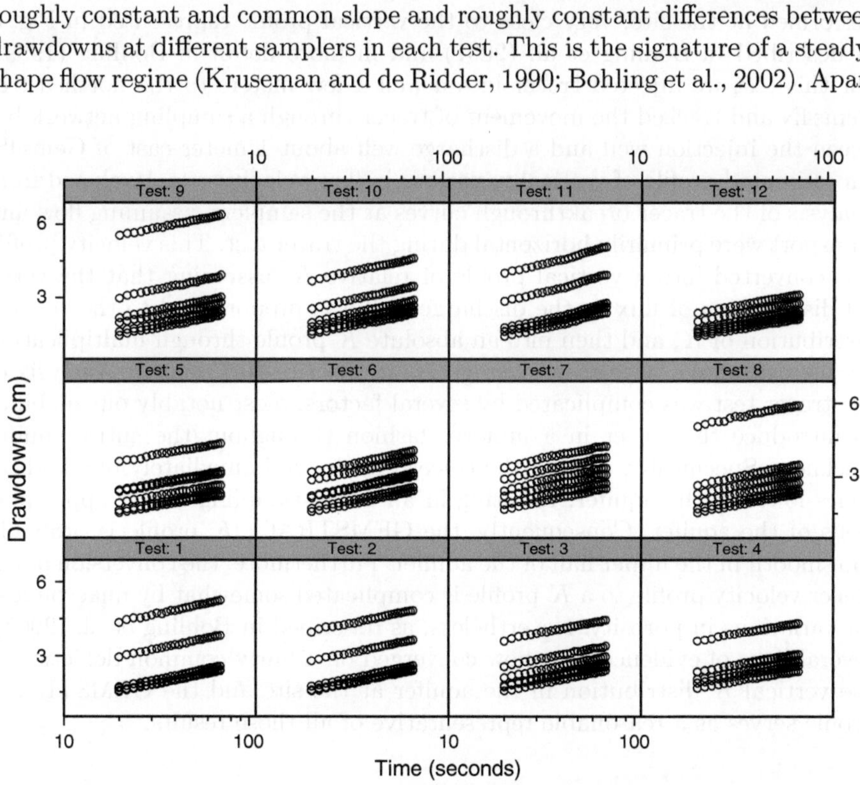

Fig. 2. Data from 12 tests in Gems4S. Test sequence is as indicated in Fig. 1. Each line of points is a sequence of observations from one of the 6 observation points used in each test.

from some changes in slope that we attribute to interference from variations in pumping at neighboring high-capacity wells, the drawdown data continue to exhibit the same steady-shape behavior (uniformly increasing drawdown and constant drawdown differences) throughout the 900 s pumping interval for each test. The next section discusses how we can take advantage of the steady-shape flow regime to reduce the computational effort involved in analyzing these data.

For all of the parameter estimation runs described in this paper we have analyzed the 20–70 second data, at two-second intervals, from the six observation ports measured in each of the 23 tests, a total of 3588 drawdown observations. Two motivations for selecting this particular time interval are that earlier data exhibit oscillations due to inertial effects (Butler and Zhan, 2004) and later data in some tests are influenced by changes in pumping at the neighboring high-capacity wells, mentioned above.

As a point of reference for the K distributions developed in this study, we use a K profile developed from an induced gradient tracer test, GEMSTRAC1, performed at the site, very close to the vertical profile represented in Fig. 1, as described in Bohling et al. (2007) and in more detail in Bohling (1999). For this test we injected tracer into a well a few meters to the northeast of Gems4N and tracked the movement of tracer through a sampling network between the injection well and a discharge well about 1 meter east of Gems4S. An estimated profile of vertical variations in flow velocity was developed from analysis of the tracer breakthrough curves at the samplers, assuming flow and transport were primarily horizontal during the tracer test. This velocity profile was converted into a vertical profile of relative K, assuming that the vertical distribution of flux to the discharge well was proportional to the vertical distribution of K, and then into an absolute K profile through multiplication by the estimated large-scale average K at the site (0.15 cm/s). Analysis of the tracer test was complicated by several factors, most notably our inability to introduce the tracer in a uniform fashion throughout the entire aquifer thickness. Specifically, most of the tracer was drawn immediately into high-K zones lower in the aquifer, resulting in an under-sampling of the upper portions of the aquifer. Consequently, the GEMSTRAC1 K profile is probably too smooth in the upper half of the aquifer. Furthermore, the conversion of the tracer velocity profile to a K profile is complicated somewhat by unaccounted for variations in porosity. Nevertheless, as described in Bohling et al. (2007), several lines of evidence have now converged on a fairly common depiction of the vertical K distribution in the aquifer at the site and the GEMSTRAC1 profile serves as a reasonable representative of all those results.

3 Analysis Methodology

Flow to a partially-penetrating pumping well in a confined aquifer is governed by the radial flow equation derived from Darcy's law and conservation of mass

(Bohling and Butler, 2001):

$$\frac{1}{r}\frac{\partial}{\partial r}\left(rK_r\frac{\partial h}{\partial r}\right)+\frac{\partial}{\partial z}\left(K_z\frac{\partial h}{\partial z}\right)=S_S\frac{\partial h}{\partial t} \tag{1}$$

where h is the hydraulic head, r is the radial distance from the center of the pumping well, z is the vertical coordinate, K_r is the hydraulic conductivity in the horizontal (radial) direction, K_z is the hydraulic conductivity in the vertical direction, and S_s is the specific storage of the aquifer. The inner (pumping well) boundary over the pumping interval can be expressed as

$$\left(rK_r\frac{\partial h}{\partial r}\right)_{r=rw}=\frac{Q}{2\pi b} \tag{2}$$

where r_w is the radius of the pumping well, Q is the pumping rate, and b is the vertical thickness of the pumping interval. In the simplest case, the left-hand side of Equation 2 is zero for z values outside the pumping interval. The numerical model employed in this study actually incorporates a representation of the wellbore, including the packers isolating the pumping interval, into the model grid, allowing for simulation of delayed responses due to wellbore storage (in the transient case) and the impact of the presence of a highly conductive wellbore. This complicates the expression of the inner boundary condition, but is somewhat immaterial to the discussion here. In this study, the upper and lower boundaries of the aquifer are represented as zero-flux boundaries and the outer boundary, placed a very large radial distance from the pumping well to minimize its impact, is a zero-drawdown boundary. In the following we will actually analyze drawdowns, rather than heads. The drawdown is given by the initial or reference head minus the head at any given point and time, so that drawdowns increase in response to pumping.

As discussed in Bohling and Butler (2001), a logarithmic transformation of the radial coordinate,

$$r'=\ln\left(r/r_w\right) \tag{3}$$

allows the radial flow equation to be recast as a two-dimensional Cartesian flow problem, and that is the approach employed in this study. For increased flexibility and convenience in analysis, we have developed a set of Matlab routines implementing a two-dimensional (radial-vertical) flow model very similar to the Fortran program described in Bohling and Butler (2001), including both transient and steady-state simulation options. For the steady-state option, the right-hand side of Equation 1 is zero and no time-stepping is involved. The code uses a standard node-centered finite difference grid in the transformed (Cartesian) space, with harmonic averages of grid cell K values used to represent cell-face conductivities. The logarithmic transformation of the radial coordinate produces a physical-space grid whose radial cell widths increase exponentially away from the pumping well. This telescoping grid in the radial direction reflects the physics of radial flow, with successively larger volumes of aquifer controlling the head response as the cone of depression expands outward (Butler, 1990). It also easily allows placement of the outer radial boundary, which is something of an artificial characteristic of the model, well away from the pumping well, reducing its impact. We have used a grid with 70 nodes with a spacing of 0.15 meters in the vertical and 60 nodes with a spacing of 0.2 along the transformed (logarithmic) radial coordinate. The model grid is 10.67 meters thick in the vertical direction, with the base (lower edge of the bottom cell) corresponding to datum ($z = 0$) in Fig. 1 and other figures. For the tests with pumping in Gems4N, the origin of the radial coordinate corresponds to the location of Gems4N and expands to the right in Fig. 1, and similarly for the tests in Gems4S, with the radial grid expanding to the left.

As discussed in Kruseman and de Ridder (1990), many confined aquifer systems rapidly reach steady-shape or transient steady state (as they call it) conditions soon after the initiation of pumping. Under these conditions, the aquifer volume within a certain radial distance of the pumping well is no longer contributing water from storage, but is simply serving as a conduit for water drawn from further away. In this case, the drawdown configuration inside that radial distance – let us say, in the region of investigation – has reached a constant shape and the gradients within the region are no longer changing. In fact, the gradients, and thus the differences in drawdown between any pair of locations, have reached their final steady-state values, even though drawdown is still increasing, in a spatially uniform fashion, over time. The system within the region of investigation can be conceptualized as a steady-state system responding instantaneously to a time-varying head boundary at some arbitrary radius. As described in Bohling et al. (2002), the attainment of steady-shape conditions allows a considerable reduction in the computational effort required for pumping test analysis: Observed differences in drawdown from different locations but common measurement times can be compared to drawdown differences extracted from a steady-state model. Bohling et al. (2007) demonstrate that steady-shape conditions are obtained within about 20–30 seconds at all observation locations for the GEMS tomographic pumping tests and that steady-shape analyses yield very similar K estimates to fully transient analyses in a fraction of the computing time.

The steady-shape approach is used for all of the analyses presented here. Each test involves 26 observation times and six observation locations. For each observation time, the observed differences between the 15 possible pairs of observation locations are compared to the corresponding drawdown differences predicted by a steady-state model. Because these differences are roughly constant over time, as dictated by the steady-shape conditions, the 26 observation times essentially yield 26 repeat measurements of each pairwise difference, yielding a total of 390 observed drawdown differences for each test, and 8970 observed drawdown differences over the set of 23 tests.

The primary objective function for all the analyses employed here is therefore the sum of squared differences between observed drawdown differences and the corresponding drawdown differences predicted by a steady-state model using the current estimate of the K distribution:

$$\|\mathbf{d}_{obs} - \mathbf{d}_{prd}\|_2^2 = \sum_{i=1}^{n} (d_{i,obs} - d_{i,prd})^2 \quad (4)$$

where $d_{i,obs}$ and $d_{i,prd}$ are the ith observed and predicted drawdown differences, and the sum runs over the $n = 8970$ observations from the 23 tests. The left-hand side is the vector representation of the same sum, which will be referred to hereafter as the residual norm. With the drawdown differences expressed in centimeters, the residual norm is expressed in squared centimeters. To this objective function we add the regularization terms described in the following sections.

We have used the function *lsqnonlin*, from Matlab's optimization toolbox, to minimize the objective function in each case. This function uses a trust region-based Newton minimization algorithm described in Coleman and Li (1996). For the analyses described here, we have provided the *lsqnonlin* function with a handle to a function returning the vector of residuals between observed and predicted drawdown differences, augmented by the "model residual" components representing the regularization terms. In the absence of a function handle returning the Jacobian (sensitivity) matrix for the problem, *lsqnonlin* uses a finite-difference approximation to the Jacobian.

The fitting parameters in all cases are actually expressed as a vector of deviations of the natural logarithm of hydraulic conductivity K, in cm/s, from a reference model, that is as

$$Y' = Y - Y_0 = \ln K - \ln K_0 \quad (5)$$

where K_0 represents the reference model. For the zeroth-order Tikhonov regularization, the regularization terms added to the objective function represent the sum of squared values of Y', referred to as a the model norm, multiplied by a regularization parameter.

Even before adding the regularization terms we restrict the parameter space for the inversion significantly by imposing a layered zonation on the model: The 70 cells in the vertical direction are partitioned into 35 layers, each 0.3 meters (two grid cells) in thickness, so that the parameter vector consists of the Y' values associated with these 35 layers. While restricting the investigation to a characterization of the K field in terms of strictly vertical variation may seem relatively modest compared to the aims of tomography, this represents a logical first step in analyzing these tests for several reasons. First, within the lateral distance (about 10 meters) investigated by these tests, we would expect vertical variation to be the predominant aspect of heterogeneity in the K field. Secondly, as in geophysical tomography, we expect this test configuration to be primarily sensitive to vertical variation. Finally, the geophysical data that we have available consist of vertical profiles of radar velocity and attenuation from a zero-offset radar profile run between Gems4N and Gems4S, so that a layered representation of the K field allows a more natural means to exploit that geophysical information. Analysis of the tomographic pumping test using a less restrictive representation of the K field is underway.

It is to be noted that the steady-shape analysis does not involve the aquifer specific storage, S_s. In fact, the establishment of steady-shape conditions dictates that the drawdown differences in the region of investigation are no longer sensitive to the storage properties of that region. In addition, we have assumed that $K_z = K_r$ since we have not seen evidence of significant anisotropy at this site.

In the next section we investigate addition of a regularization term to the inversion problem to incorporate *a priori* information on the K model, represented in terms of a uniform background model, and then incorporate the available geophysical information through representation of the estimated $Y = \ln K$ values as an arbitrary smooth function of the geophysical parameters, in Sect. 5.

4 Regularization Relative to Uniform Model

In this section we apply zeroth-order Tikhonov regularization relative to a uniform background model to the analysis of the tomographic pumping tests. Tikhonov regularization is commonly applied in geophysical tomography, as described in Aster et al. (2005). Tonkin and Doherty (2005) discuss the use of Tikhonov regularization for inverse problems in hydrogeology. Aster et al. (2005) point out that Tikhonov regularization is very much like Bayesian approaches for incorporating *a priori* model information into the inversion, but without the explicit formulation in terms of a probabilistic model.

In this analysis we have used a background model of $Y_0 = \ln K_0 = \ln(0.15\,\mathrm{cm/s}) = -1.9$ and the objective function is formulated as

$$F\left(Y'\right) = \sum_{i=1}^{n} \left(d_{i,obs} - d_{i,prd}\right)^2 + \sum_{j=1}^{35} \left(\alpha Y_j'\right)^2 = \left\|\mathbf{d}_{obs} - \mathbf{d}_{prd}\right\|_2^2 + \alpha^2 \left\|\mathbf{L}\mathbf{Y}'\right\|_2^2 \quad (6)$$

where α is the regularization parameter. The vector representation on the right expresses the model norm in terms of a matrix, $\mathbf{L}$, multiplying the vector of parameters. For zeroth-order regularization, this matrix is simply the identity matrix. For zeroth-order regularization, increasing α increases the penalty on deviations from the reference model, Y_0. Higher orders of regularization penalize deviations of the estimated parameters from a smooth model. For first-order regularization, an $\mathbf{L}$ matrix yielding first differences of Y' would be employed, penalizing variations in the (approximate) first spatial derivative of Y', whereas $\mathbf{L}$ would be the finite-difference approximation to the Laplacian, yielding second differences, for second-order regularization. Rather than simple first- and second-order differences, the regularization term could also be expressed in terms a spatial covariance model, in order to impose an expected spatial correlation structure on the estimates (Doherty, 2003).

Aster et al. (2005) recommend the use of L-curve plots to examine the tradeoff between the residual norm and the model norm with variations of the regularization parameter, α. Figure 3 shows such a plot for the zeroth-order regularized inversion of the tomographic pumping test data. For this analysis we used 16 α values ranging from 0.001 to 1.0 in equal logarithmic increments and, for each value of α, performed five different inversions starting from a different initial estimate for Y'. The initial Y' vectors were populated with zero-mean Gaussian random numbers with a small standard deviation, in order to test the sensitivity of the inversion process to small variations in the initial estimates. Thus, Fig. 3 contains five points, corresponding to the five inverse results, for each value of α. Ideally, the L-curve will show a distinct "corner", pointing towards the origin of the plot, and this corner point represents the optimal balance between data fit and model deviations (Hansen, 1992). The plot in Fig. 3 does show a reasonably well defined corner, corresponding to a residual norm of 260 cm^2 and a model norm of 46, which are obtained for $\alpha = 0.1$.

Figure 4 shows the residual norm versus α, demonstrating that the model norm ceases to decrease significantly as α decreases below 0.1. Taken together, Figs. 3 and 4 seem to indicate that 0.1 is a reasonable value for the regulariza-

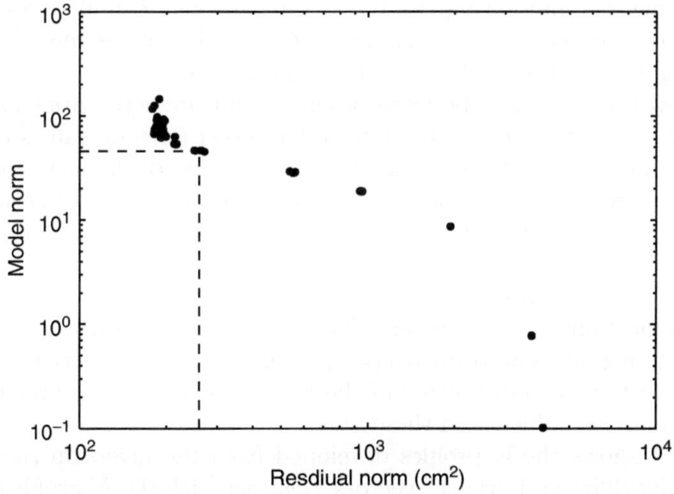

Fig. 3. L-curve for zeroth-order Tikhonov regularized inversion of tomographic pumping tests using layered zonation. Lines drawn for alpha = 0.1, with residual norm of approximately 260 cm^2 and model norm of 46 (squared deviations of lnK from reference value).

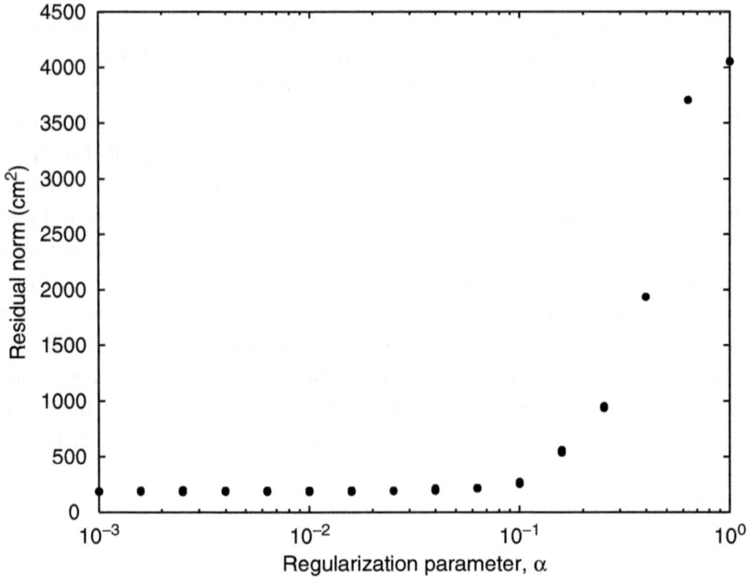

Fig. 4. Residual norm versus regularization parameter for zeroth-order Tikhonov regularized inversion of the tomographic pumping tests.

tion parameter, representing the maximum amount of model regularization that can be achieved without significant degradation of the model fit.

To illustrate the tradeoff between residual norm and model norm, Figs. 5 and 6 show the crossplots between observed and predicted drawdown differences and the corresponding K profiles developed for four values of α, 0.06, 0.10, 0.16, and 0.25, and one of the five "realizations" of the inversion, namely that starting from the third of the five initial vectors. The background model of $K = 0.15\,\text{cm/s}$ is represented by the vertical line in each panel of Fig. 6, and the damping effects with increasing regularization are clear in the plot. Note that the estimates tend to stay closer to the background value in the upper and lower portions of the aquifer. The data are less sensitive to the K values in these regions due to the decreased "coverage" of the test configuration (Fig. 1) near the top and bottom of the aquifer. Therefore, the regularization term exerts more influence in these areas.

Figure 7 shows the K profiles developed from the inversion runs starting from the five different initial Y' vectors, together with the K profile developed from the GEMSTRAC1 tracer test. These five profiles have many features in common, demonstrating that the inversion results are not overly sensitive to small variations in the initial vectors. The standard deviation of the random values in these initial vectors is about 0.1, in log space, compared to standard deviations of about 1.2 for the final Y' profiles, so that the initial "noise" is about $1/10^{\text{th}}$ the level of the final level of variability in each profile.

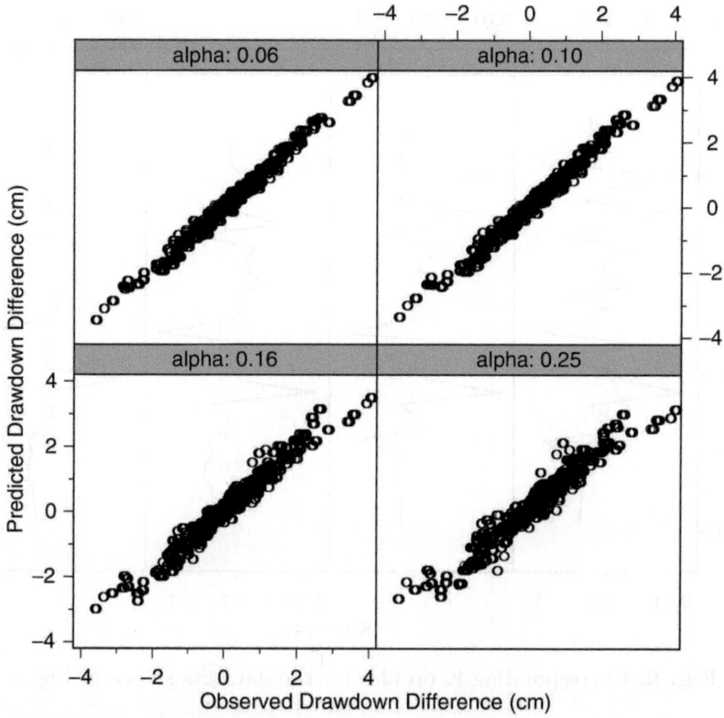

Fig. 5. Observed and predicted drawdown differences for "realization 3" using different values of the regularization parameter, α ("alpha" in the panel captions).

An important result of comparing estimated profiles for different starting vectors, and across different levels of regularization, as in Fig. 6, is to see which features appear consistently, as these can be taken as the features which are most demanded by the data. The double-peaked structure of high-K zones between about 4 and 5.5 meters above datum is reasonably consistently reproduced in the profiles in Fig. 7, and this structure probably corresponds with the double-peaked structure in the GEMSTRAC1 profile, although with a bit of a depth shift. In addition, the high-K peak just above 2 meters above datum in the GEMSTRAC1 profile seems to be matched by corresponding peaks in the tomography profiles for trials 1, 3, and 4, although this peak is obscured by extended regions of high K in the lower portions of the profiles for realizations 2 and 5.

The tomography profiles show considerably more structure in the upper portion of the aquifer than does the GEMSTRAC1 profile. In fact, the GEMSTRAC1 tracer test did not sample the upper portion of the aquifer adequately, due to the fact that most of the tracer was drawn into the high-K zones lower in the aquifer. Therefore, the upper portion of the GEMSTRAC1 K profile is probably smoother than it should be. Another reason for the

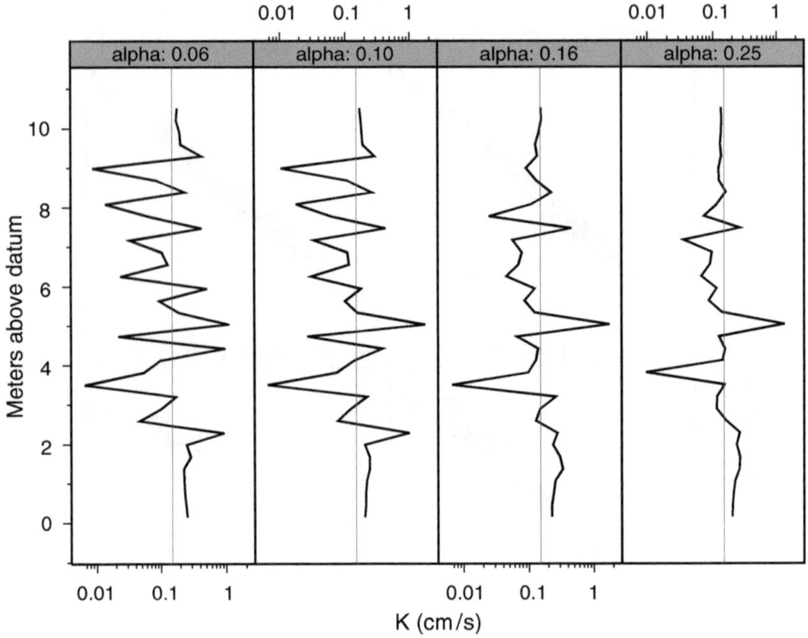

Fig. 6. Corresponding K profiles for the data fits shown in Fig. 5.

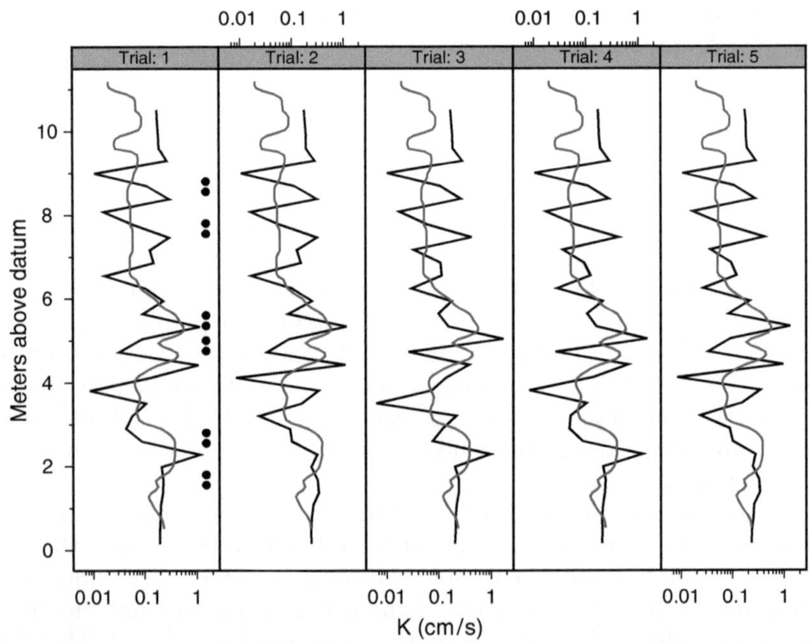

Fig. 7. Profiles of five different zeroth-order Tikhonov regularized inversion results for $\alpha = 0.1$, together with the K profile developed from the GEMSTRAC1 tracer test (gray curve).

difference in character between the GEMSTRAC1 K profile and the tomography profiles is that the GEMSTRAC1 profile was essentially developed in "arithmetic K" space, rather than "log K" space, and consequently does not show the same level of variability at the low K end as is exhibited in the tomography profiles.

As discussed in Doherty (2003), the reference model, Y_0, could be generated in a number of ways and needn't be a simple uniform model. For example, the fitted Y' values could represent deviations from a model generated through stochastic simulation or kriging from existing data.

5 Regularization Relative to Geophysical Attributes

Figure 8 shows the K profile derived from the GEMSTRAC1 tracer test together with vertical profiles of radar velocity and attenuation derived from a zero-offset survey between Gems4N and Gems4S. Bohling et al. (2007) describe the acquisition of the radar data in more detail. In that work we employed the radar data to assist in the zonation of the flow model employed in analyzing the tomographic pumping tests, on the presumption that the electromagnetic and hydraulic properties of the aquifer sediments are governed to some extent by the same underlying lithologic variations. Based on cluster

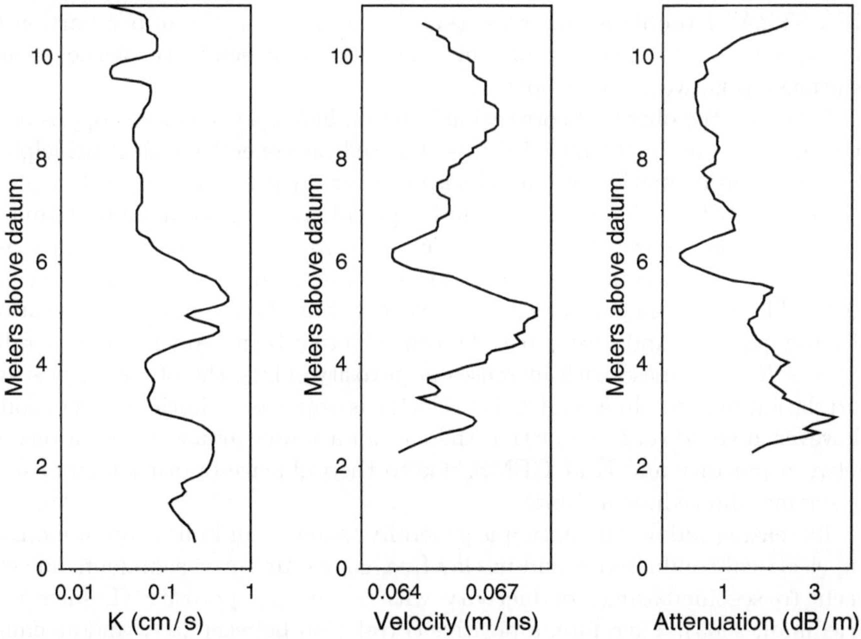

Fig. 8. GEMSTRAC1 K profile compared to velocity and attenuation profiles obtained from a zero-offset radar survey between Gems4N and Gems4S.

analysis of the radar velocity and attenuation data, we defined four more coarsely layered zonations, with five, seven, ten, and thirteen layers, and estimated K values for those zones in the pumping test analysis. Comparisons to K profiles developed using equal-thickness zonations with comparable numbers of layers and with other hydraulic test data from the site seems to indicate that the radar data does indeed contain useful information regarding the hydraulic structure of the aquifer.

Here we will attempt to exploit more direct quantitative correlations that might exist between K and the radar properties by incorporating the velocity and attenuation profiles into our estimation of the 35-layer K profile through a regularization term specifying our desire for the estimated $Y = \ln K$ profile to be nicely represented as a smooth function of the radar attributes. Prior to doing so, we should ask ourselves what relationships we might expect to see between these properties and what relationships, if any, seem to be indicated by the available data. A glance a Fig. 8 gives the impression of a positive correlation between the GEMSTRAC1 $\ln K$ profile and both of the radar property profiles, at least in the lower half of the aquifer. In fact, the collocated profiles are not as strongly correlated as one might guess from Fig. 8. Considering the entire sequence where radar data are available, from 2.3 to 10.7 meters above datum, there is a linear correlation of 0.16 between $\ln K$ and radar velocity and 0.44 between $\ln K$ and attenuation. Considering only data below 7 meters, the correlations change to 0.41 between $\ln K$ and velocity and 0.21 between $\ln K$ and attenuation. As mentioned above, it is possible that the GEMSTRAC1 profile is smoother than it should be in the upper portion of the aquifer, so the latter results may be more representative of the "true" relationship between the properties.

However, the observed correlation between $\ln K$ and velocity is opposite in sign to what one might initially expect based on general physical principles. As a first approximation, radar velocity is inversely proportional to the square root of the dielectric constant or relative permittivity of the medium (Annan, 2005) and the square root of the dielectric constant is expected to increase in proportion to the porosity in a water-saturated medium (Wharton et al., 1980). Therefore, higher velocities would generally be taken as an indication of lower porosity and vice-versa. All other factors being equal, one generally expects K to increase with increases in porosity. Thus, the observed positive correlation between $\ln K$ and radar velocity is opposite to initial expectations. However, it could very well be that there is not a simple monotonic relationship between porosity and K at GEMS, due to the influence of other factors such as sorting and sediment fabric.

Increasing radar attenuation is generally taken as an indication of increasing electrical conductivity and in clay-free, coarse-textured sediments one expects to see increasing conductivity with increasing porosity (Lesmes and Friedman, 2005). Therefore, a positive correlation between $\ln K$ and attenuation is not unexpected.

To incorporate the radar velocity and attenuation into the $\ln K$ estimation without needing to specify the exact form of the relationship between these properties in advance, we have added a regularization term that penalizes deviations between the current estimate of the $Y = \ln K$ profile and that predicted from radar attributes using a smoothing spline model. We also retain the regularization relative to a uniform background model, Y_0, with the corresponding regularization parameter fixed at $\alpha_0 = 0.1$, as developed in the previous section. It is quite possible that the optimal value of α_0 would change with the addition of the regularization term relative to geophysical properties, but for simplicity we do not explore that option here. The complete objective function is now given by

$$F(Y') = \|\mathbf{d}_{obs} - \mathbf{d}_{prd}\|_2^2 + \alpha_0^2 \left\|\mathbf{L}\mathbf{Y}'\right\|_2^2 + \alpha^2 \left\|(\mathbf{Y}' + \mathbf{Y}_0) - f(v, \gamma; \mathbf{Y}' + \mathbf{Y}_0)\right\|_2^2 \tag{7}$$

where α now represents the regularization parameter relative to the geophysical parameters and $f(v, \gamma; \mathbf{Y}' + \mathbf{Y}_0)$ represents the smoothing spline model approximation of $\mathbf{Y} = \mathbf{Y}' + \mathbf{Y}_0$ based on the radar velocity and attenuation, v and γ. It is important to emphasize that this smoothing spline model relates the current estimate of $\mathbf{Y}$, as the response variable, to the radar attributes, as predictor variables, and is therefore re-estimated at each step of the inversion process. The spline function does not depend in any way on "hard" K data or previously existing estimates of K (such as those from GEMSTRAC1), but is instead developed in an iterative fashion as part of the inversion process.

In this work we have used the thin-plate smoothing spline implemented in the *tpaps* function in Matlab's spline toolbox. As described in Hastie et al. (2001), a thin-plate spline builds an approximation of the response variable (Y in this case) from a set of radial basis functions centered on the predictor variable (v and γ) data points, with a smoothing parameter chosen to optimize the balance between model fit and model smoothness, measured in terms of the second derivative of the modeled response function. In this study, the predictor variables, v and γ, represent standardized versions of the radar velocity and attenuation, each scaled to zero mean and unit standard deviation to equalize their weight in the developed model. The model relates the estimated $\mathbf{Y}$ values to the standardized radar attributes, v and γ, for the 28 model layers encompassed by the radar profiles and is developed to minimize the following residual sum of squares:

$$RSS(f, p) = p \sum_{i=1}^{28} \{Y_i - f(v_i, \gamma_i)\}^2 + (1 - p) R(f) \tag{8}$$

where $R(f)$ is the integral of the second derivative of $f(v, \gamma)$. The *tpaps* function uses an ad-hoc procedure for estimating the smoothing parameter, p, based on the distribution of data points in predictor variable space. In this particular case, the selected smoothing parameter is $p = 0.23$.

Other nonparametric smoothing models, such as neural networks or locally-weighted regression (Hastie et al., 2001) could also be used in place of a smoothing spline and the response function could easily incorporate multiple variables, if additional geophysical (or other) attributes were available. In addition, we have used a smoothing function relating $\ln K$ only to the collocated values of radar velocity and attenuation, interpolating the radar profiles to the centers of the 28 model layers covered by the geophysical profiles (with the seven K zones at the bottom of the aquifer remaining "unregularized"). The smoothing function could also be formulated to incorporate spatially lagged or averaged geophysical attributes if that were deemed appropriate.

Regardless of the exact nature of the smoothing function, increasing the regularization parameter, α, will force the estimated Y values to more closely match the characteristics of the geophysical attributes. Because the form of the relationship is not specified in advance of the inversion, an *a posteriori* assessment of its plausibility, after the inversion is complete, is required.

We have run a series of inversions using the objective function expressed in Equation 7, with 16 α values again ranging from 0.001 to 1.0 in equal log increments, and with the background model regularization parameter, α_0, fixed at 0.1. As before, we have run five different inversions, starting from the same set of five initial Y' vectors, for each value of α. Figure 9 shows the trade-off between the drawdown difference residual norm and the spline fit residual norm for these runs and Fig. 10 shows the plot of the drawdown difference residual norm versus α. In this case, as α increases, placing more weight on the third term in Equation 7, the developed $\ln K$ profile will more

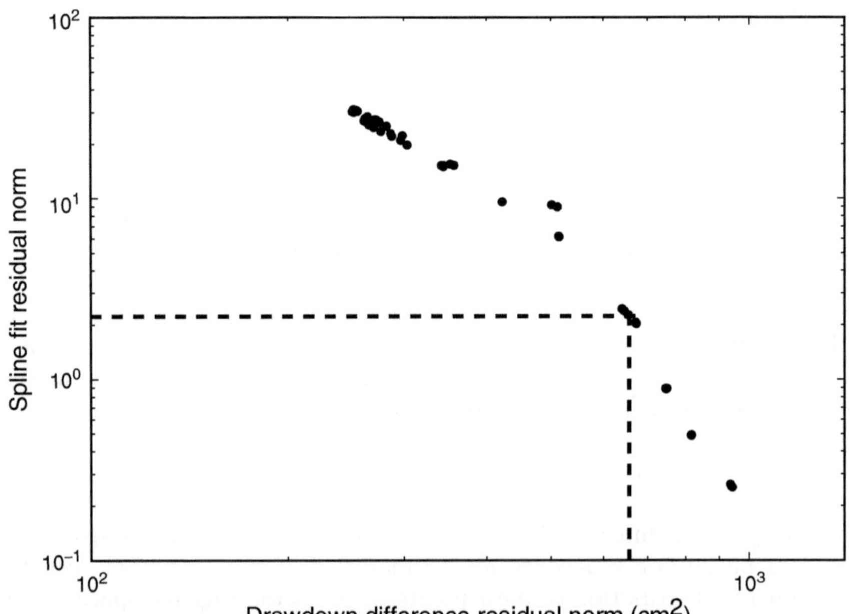

Fig. 9. "L-curve" for regularization relative to spline function of geophysical parameters. The lines indicate the results for $\alpha = 0.25$, with a drawdown difference residual norm of 657 cm^2 and a spline fit residual norm of 2.23.

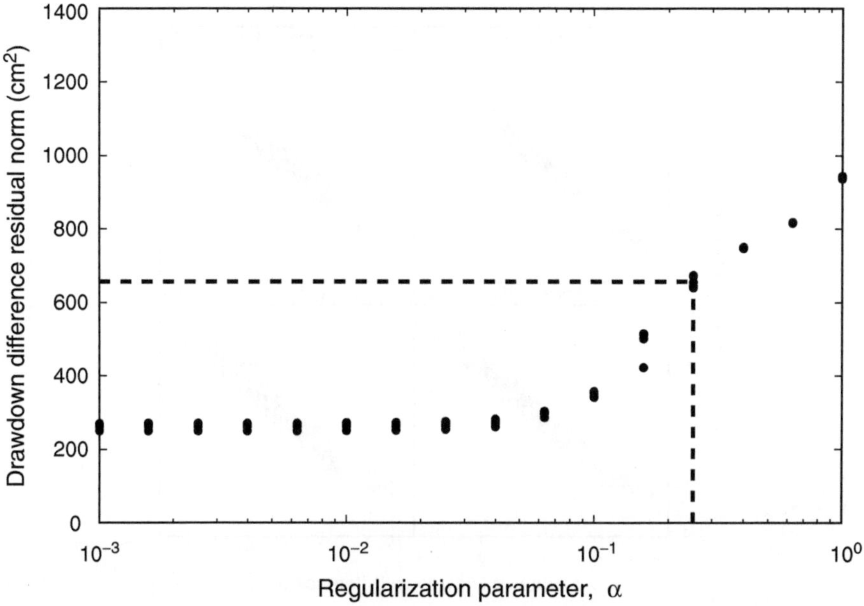

Fig. 10. Drawdown difference residual norm versus regularization parameter for regularization relative to geophysical parameters.

strongly reflect the radar attributes, reducing the spline fit residual norm and (most likely) increasing the sum squared residual between observed drawdown differences and those precdicted by the flow model. Unfortunately, as shown in Fig. 9, this tradeoff does not follow a clear L shape in this case, making it somewhat difficult to select an appropriate value of α. For the sake of demonstrating the ability of the proposed technique to imbue the Y estimates with the characteristics of the geophysical data, we will focus on the results for $\alpha = 0.25$, even though this is probably too strong a regularization. The five results for $\alpha = 0.25$ on Fig. 9 show an average drawdown difference residual norm of 657 cm^2 and a spline fit residual norm of 2.23. It is clear from Fig. 10 that a choice of $\alpha = 0.25$ represents a fair amount of degradation in data fit relative to smaller values of α, meaning we have selected a value that is fairly strongly weighted towards regularization.

Figures 11 and 12 examine the tradeoff between data fit and regularization in detail, again for the inversions starting from the third of five different initial Y' vectors. The degradation in data fit and the increasing conformance of the K profile to the geophysical parameters with increasing α is quite clear in these figures. Figure 13 compares the K profile (on a log scale) for $\alpha = 0.25$ to the profiles of the radar velocity and attenuation, demonstrating that the regularization has strongly imbued the estimated K profile with the character of the geophysical attributes, especially the radar velocity. Indeed, one might say that this log K profile looks too much like the velocity profile and that

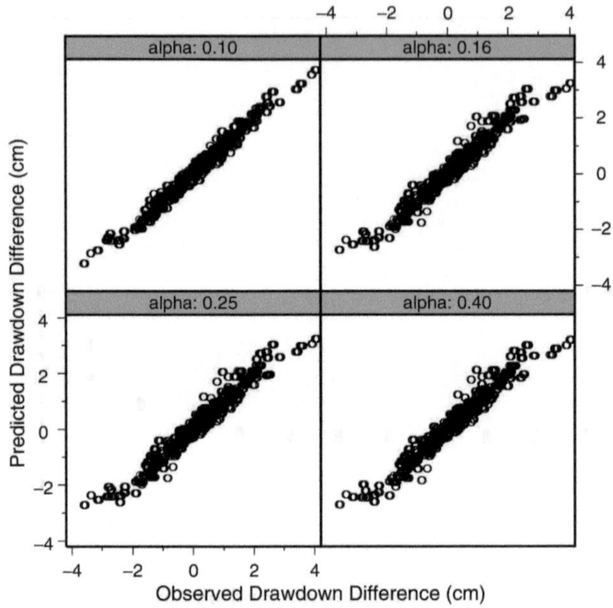

Fig. 11. Observed and predicted drawdown differences for different levels of regularization relative to geophysical attributes.

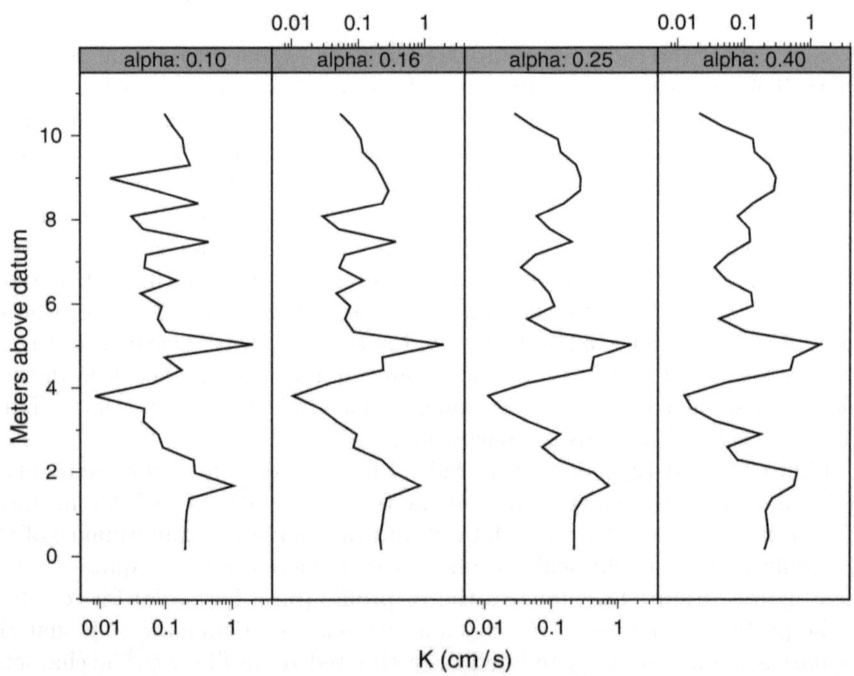

Fig. 12. Estimated K profiles for different levels of regularization, for trial three of the five inversions from different starting vectors.

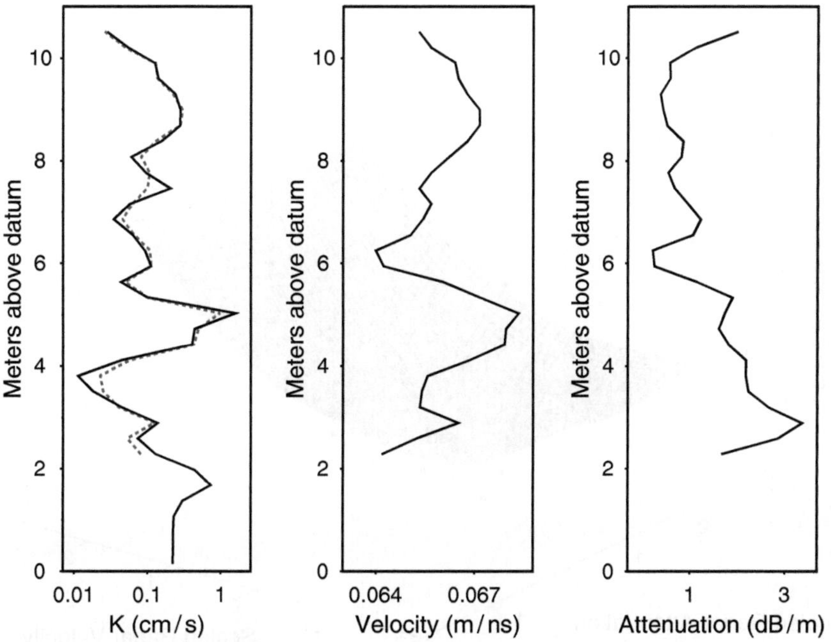

Fig. 13. Estimated K profile (solid line, left panel) for trial 3 with $\alpha = 0.25$ together with spline model prediction of K profile (gray dashed line, left panel) and radar velocity and attenuation profiles (next two panels).

a lesser degree of regularization might be appropriate. The high-K zone near the top of the aquifer, where the tomogrpahic pumping test sampling is less dense, is inherited directly from the velocity profile. The log K profile shown in Fig. 13 shows a fairly strong linear correlation, 0.60, with the radar velocity and a moderate negative correlation, −0.25, with the radar attenuation. Thus, both of these relationships are opposite to initial physical expectations, as described above. Nevertheless, as discussed in Hyndman and Tronicke (2005), relationships between hydrologic and geophysical parameters are both and uncertain and non-unique, and therefore need to be developed on a site-specific basis. Additional investigation would be required to determine the plausibility of the relationships developed here, but both the GEMSTRAC1 profile and the results presented in Bohling et al. (2007) indicate the presence of a positive correlation between K and radar velocity at this site. In contrast, the log K profile developed using a regularization parameter of 0.1 (first panel of Fig. 12) exhibits a correlation of 0.27 with the radar velocity, more in keeping with that displayed by the GEMSTRAC1 profile, and essentially no correlation with the radar attenuation. This may in fact represent a more reasonable level of regularziation.

The dashed gray line in the left panel of Fig. 13 represents the spline model prediction of log K as a function of the radar attributes. Again, it is important

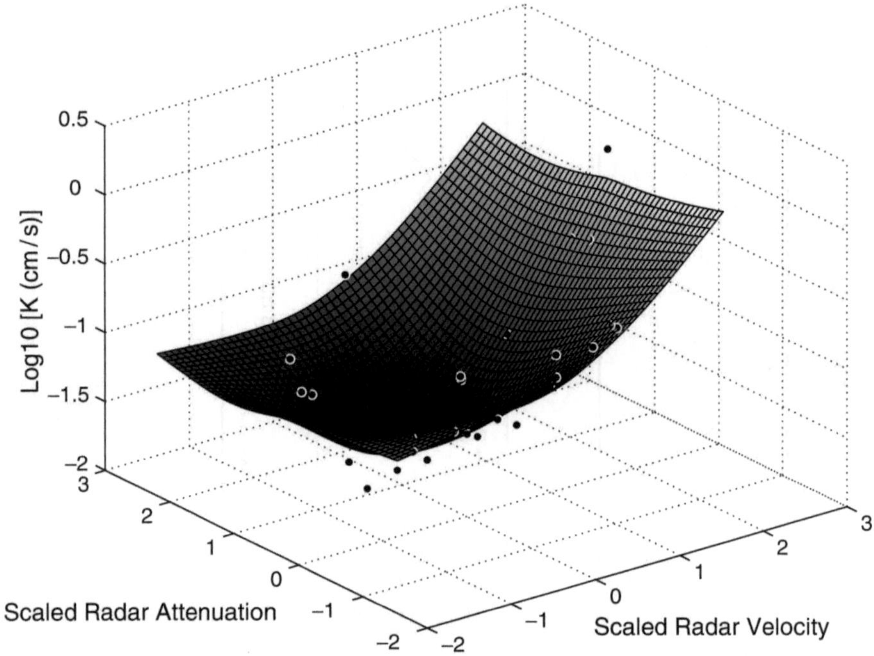

Fig. 14. Smoothing spline relationship developed between log K and radar attributes for K field estimated using $\alpha = 0.25$, trial 3.

to keep in mind that the spline model relates the log K values estimated by the inversion (solid line in the left panel of Fig. 13) to the radar attributes and that the inversion has been strongly encouraged to produce a log K distribution that can be nicely represented as a spline function of the radar attributes. The differences between the spline model representation of K and the K values estimated by the inversion are dictated by the other two terms in the objective function – namely the data misfit and the deviations from a uniform background log K model. Figure 14 shows the smoothing spline model itself. The points in this plot represent the log K estimates from the inversion process (those represented by the solid line in the left pane of Fig. 13) plotted versus the corresponding values of scaled radar velocity and attenuation for the 28 model layers with radar data. The surface represents the resulting smoothing spline model. This function is nonlinear and non-monotonic along both predictor variable axes, demonstrating the ability of the proposed technique to accommodate fairly arbitrary relationships between hydrologic and geophysical parameters.

Figure 15 shows the log K profiles developed from the five different Y' starting vectors using $\alpha = 0.25$. Not surprisingly, all five profiles look quite a bit like the radar velocity profile, demonstrating the reduction in variability

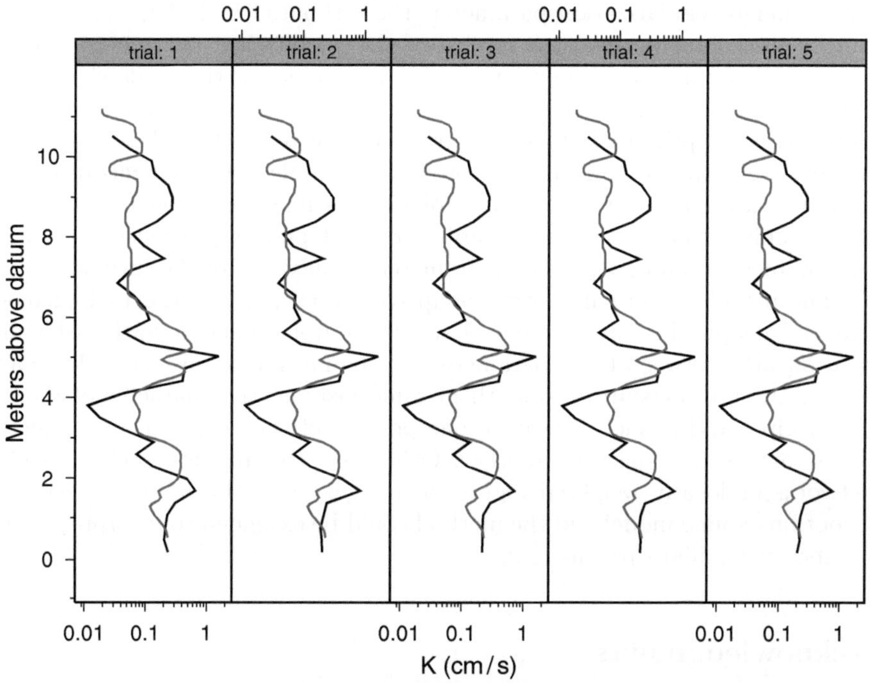

Fig. 15. Five inversion results using $\alpha = 0.25$ for regularization relative to the geophysical parameters, along with the GEMSTRAC1 K profile.

that one would hope to obtain from incorporation of geophysical data, at the cost of imposing perhaps too strong a dependence on that data in this case.

6 Conclusions

In this study we have investigated a simple approach for fusing hydrogeological and geophysical information by direct inclusion of the geophysical information in a regularization term representing our desire for the hydrogeological parameters estimated in an inverse analysis to reflect the spatial character of geophysical parameters. Rather than specifying the functional form of the relationship in advance, we have used a smoothing spline model to represent the relationship between the hydrogeological parameter, in this case the logarithm of hydraulic conductivity, $\ln K$, and the geophysical parameters, radar velocity and attenuation. At each step of the hydrologic inversion, the smoothing spline model sees the current estimate of $\ln K$ as the response variable to be modeled as a function of velocity and attenuation and thus the inclusion of this regularization term tends to drive the developing $\ln K$ profile towards one that can be represented as a smooth function of the geophysical parameters, the more so as the regularization parameter increases. In fact, with a large enough regularization parameter, the estimated hydrologic parameters can be forced to be almost a replica of the geophysical parameters, meaning that this approach requires a careful assessment of the tradeoff between data fit and regularization and of the plausibility of the developed relationships, based on prior expectations derived from available field data or physical models. Nevertheless, this method provides a simple means for incorporating geophysical information into hydrogeological estimation problems, especially where limited field data are available for developing a relationship between hydrogeological and geophysical parameters in advance of the inversion.

The use of a non-parametric, adaptive model relating the hydrogeological and geophysical parameters allows the development of fairly arbitrary relationships between these parameters as the inversion proceeds. All we ask is that the relationship be smooth in some fashion; the character of the relationship could in fact vary from one portion of the geophysical parameter predictor variable space to another. Other adaptive models, such as neural networks or locally weighted regression models, could be used in place of a smoothing spline model and the method could be extended to incorporate any number of auxiliary parameters.

Acknowledgments

This research was supported by the Hydrologic Sciences Program of the National Science Foundation under grant 9903103 and the Kansas Geological Survey Applied Geohydrology Summer Research Assistantship Program. Any opinions, findings and conclusions, or recommendations expressed in this paper are those of the authors and do not necessarily reflect the views of the NSF. The field efforts were led by Dr. James J. Butler, Jr., of the Kansas Geological Survey, with assistance provided by John Healey of the Kansas Geological Survey and Greg Davis and Sam Cain, participants in the Applied Geohydrology Summer Research Assistantship Program. We thank the two anonymous reviewers for their helpful comments.

References

Annan AP (2005) GPR methods for hydrogeological studies. In: Rubin Y, Hubbard SS (eds) Hydrogeophysics. Springer, Dordrecht, pp 185–213

Aster RC, Borchers B, Thurber CH (2005) Parameter estimation and inverse problems. Elsevier Academic Press, Amsterdam

Bohling GC (1993) Hydraulic tomography in two-dimensional, steady-state groundwater flow. Eos Trans. AGU 74, 141

Bohling GC (1999) Evaluation of an induced gradient tracer test in an alluvial aquifer. Ph.D. thesis, University of Kansas

Bohling GC, Butler JJ Jr. (2001) Lr2dinv: A finite-difference model for inverse analysis of two-dimensional linear or radial groundwater flow. Computers & Geosciences 27:1147–1156

Bohling GC, Butler JJ Jr, Zhan X, Knoll, MD (2007) A field assessment of the value of steady-shape hydraulic tomography for characterization of aquifer heterogeneities. Water Resour. Res. 43, W05430, doi:10.1029/2006WR004932.

Bohling GC, Zhan X, Butler JJ Jr., Zheng L. (2002) Steady-shape analysis of tomographic pumping tests for characterization of aquifer heterogeneities. Water Resour. Res., 38, doi:10.1029/2001WRR001176

Brauchler R, Liedl R, Dietrich P (2003). A travel time based hydraulic tomographic approach. Water Resour. Res. 39, doi:10.1029/2003WR002262

Butler JJ Jr. (1990) The role of pumping tests in site characterization: Some theoretical considerations. Ground Water 28:394–402

Butler JJ Jr. (2005) Hydrogeological methods for estimation of hydraulic conductivity. In: Rubin Y, Hubbard SS (eds), Hydrogeophysics. Springer, The Netherlands, pp. 23–58

Butler JJ Jr., McElwee CD, Bohling GC (1999) Pumping tests in networks of multilevel sampling wells: Motivation and methodology. Water Resour. Res. 35:3553–3560

Butler JJ Jr., Zhan X (2004), Hydraulic tests in highly permeable aquifers, Water Resour. Res., 40, W12402, doi:10.1029/2003WR002998.

Carrera J, Neuman SP (1986), Estimation of aquifer parameters under transient and steady state conditions: 1. Maximum likelihood method incorporating prior information Water Resour. Res. 22:199–210

Carrera J, Alcolea A, Medina A, Hidalgo J, Slooten LJ (2005), Inverse problems in hydrogeology. Hydrogeology Journal 13:206–222

Coleman TF, Li Y (1996) An interior, trust region approach for nonlinear minimization subject to bounds. SIAM Journal on Optimization 6:418–445

Cooper HH Jr, Jacob CE (1946). A generalized graphical method for evaluating formation constants and summarizing well field history. Trans. AGU, 27:526–534

Doherty J (2003) Ground water model calibration using pilot points and regularization. Ground Water 41:170–177

Freeze RA, Cherry JA (1979) Groundwater. Prentice-Hall, Inc., Englewood Cliffs, NJ.

Gottlieb J, Dietrich P (1995) Identification of the permeability distribution in soil by hydraulic tomography. Inverse Probl. 11:353–360

Hansen PC (1992) Analysis of discrete ill-posed problems by means of the L-curve. SIAM Review 34:561–580

Hastie T, Tibshirani R, Friedman J (2001) The elements of statistical learning: Data mining, inference, and prediction. Springer, New York

Hyndman D, Tronicke J (2005) Hydrogeophysical case studies at the local scale: the saturated zone. In: Rubin Y, Hubbard SS (eds) Hydrogeophysics. Springer, Dordrecht, pp 391–412

Kruseman GP, de Ridder NA (1990) Analysis and Evaluation of Pumping Test Data – ILRI Pub. 47. Int. Inst. for Land Reclamation and Improvement, the Netherlands

Lesmes DP, Friedman SP (2005) Relationships between the electrical and hydrogeological properties of rocks and soils. In: Rubin Y, Hubbard SS (eds) Hydrogeophysics. Springer, Dordrecht, pp 87–128

Liu S, Yeh T-CJ, Gardiner R (2002) Effectiveness of hydraulic tomography: Sandbox experiments, Water Resour. Res. 38, doi: 10.1029/2001WR000338

Neuman SP (1987) Stochastic continuum representation of fractured rock permeability as an alternative to the REV and fracture network concepts. In: Proc. U.S. Symp. Rock Mech., 28P[thP], pp. 533–561

Parker RL (1994) Geophysical Inverse Theory. Princeton University Press

Rubin Y, Hubbard SS (eds) (2005) Hydrogeophysics. Springer, Dordrecht

Tonkin MJ, Doherty J (2005) A hybrid regularized inversion methodology for highly parameterized environmental models. Water Resources Research 41, doi:10.1029/2005WR003995

Tosaka H, Masumoto K, Kojima K (1993) Hydropulse tomography for identifying 3-d permeability distribution. In: High Level Radioactive Waste Management: Proceedings of the Fourth Annual International Conference of the ASCE, Am. Soc. Civ. Eng., Reston, Va, pp. 955–959

Tronicke J, Holliger K, Barrash W, Knoll MD (2004) Multivariate analysis of cross-hole georadar velocity and attenuation tomograms for aquifer zonation, Water Resour. Res., 40, doi:10.1029/2003WR002031

Vesselinov VV, Neuman SP, Illman WA (2001a) Three-dimensional numerical inversion of pneumatic cross-hole tests in unsaturated fractured tuff: 1. Methodology and borehole effects. Water Resour. Res. 37:3001–3017

Vesselinov VV, Neuman SP, Illman WA (2001b) Three-dimensional numerical inversion of pneumatic cross-hole tests in unsaturated fractured tuff: 2. Equivalent parameters, high-resolution stochastic imaging and scale effects, Water Resour. Res. 37:3019–3041

Wharton RP, Hazen GA, Rau RN, Best DL, Advancements in electromagnetic propagation logging. Society Petroleum Engineering, Paper 9041

Yeh T-CJ, Liu S (2000) Hydraulic tomography: Development of a new aquifer test method. Water Resour. Res. 36:2095–2105

Zhu J, Yeh T-CJ (2005) Characterization of aquifer heterogeneity using transient hydraulic tomography. Water Resour. Res. 41, doi:10.1029/2004WR003790

Advancing the Use of Satellite Rainfall Datasets for Flood Prediction in Ungauged Basins: The Role of Scale, Hydrologic Process Controls and the Global Precipitation Measurement Mission

Faisal Hossain and Nitin Katiyar

Department of Civil and Environmental Engineering, Tennessee Technological University

1 Introduction

Floods account for about 15% of the total death toll related to natural disasters, wherein typically more than 10 million lives are either displaced or lost each year internationally (Hossain, 2006). Rainfall is the primary determinant of floods and its intimate interaction with the landform (i.e., topography, vegetation and channel network) magnified by highly wet antecedent conditions leads to catastrophic flooding in medium (i.e., $1000 \sim 5000\,km^2$) and large (i.e., $>5000\,km^2$) river basins. Furthermore, floods are more destructive over tropical river basins that lack adequate surface stations necessary for real-time rainfall monitoring – i.e., the ungauged river basins (Hossain and Katiyar, 2006) (see Figure 1, left panel).

However, flood prediction is becoming ever more challenging in these medium-to-large river basins due to the systematic decline of in situ rainfall networks world-wide. The gradual erosion of these conventional rainfall data sources has lately been recognized as a major concern for advancing hydrologic monitoring, especially in basins that are ungauged or already sparsely instrumented (Stokstad, 1999; Shikhlomanov et al., 2002). As a collective response, the hydrologic community has recently established partnerships for the development of space-borne missions for cost-effective, yet global, hydrologic measurements. The most pertinent example in the context of flood prediction is the Global Precipitation Measurement (GPM) mission for global monitoring of rainfall (Smith et al., 2007). Hence, there is no doubt that the hydrologic community as a whole will gradually become dependent on GPM for a substantial part of its rainfall data needs for hydrologic research and operational monitoring.

F. Hossain and N. Katiyar: *Advancing the Use of Satellite Rainfall Datasets for Flood Prediction in Ungauged Basins: The Role of Scale, Hydrologic Process Controls and the Global Precipitation Measurement Mission*, Studies in Computational Intelligence (SCI) **79**, 163–181 (2008)
www.springerlink.com

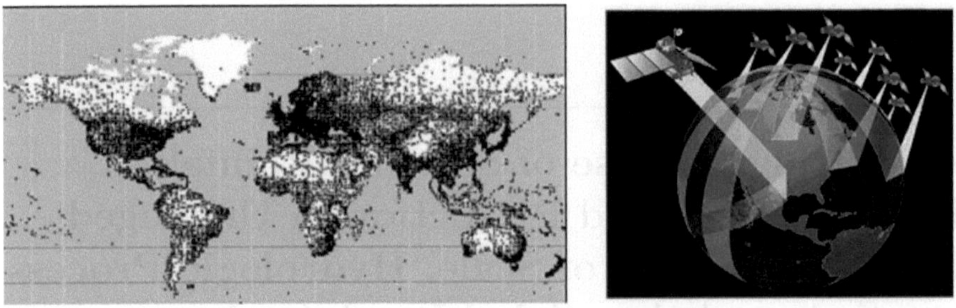

Fig. 1. Left panel – global distribution of in-situ rainfall gages showing the sparse and unevenness in the underdeveloped world (source: http://www.cpc.noaa.gov). Right Panel – Constellation of anticipated GPM satellites. The larger satellite on the left represents the core with a radar on board, while the rest carry polar orbiting PMW sensors (source: http://gpm.gsfc.nasa.gov).

GPM now beckons hydrologists as an opportunity to improve flood prediction capability in ungauged basins. However, before the potential of GPM can be realized, there are a number of hydrologic issues that must be addressed. Our success in leveraging the GPM to improve flood prediction will depend largely on the recognition of these issues and the feedback provided by hydrologists on the assessment of satellite rainfall data to the satellite data producing community (Hossain and Lettenmaier, 2006). The purpose of this chapter is to articulate these hydrologic issues that require further research and highlight the recent progress made in understanding them in the hope that satellite rainfall data can be used in hydrologic models more effectively in future.

2 Overview of Satellite Rainfall Remote Sensing and GPM

The heritage of GPM originated two decades ago when Infrared (IR) radiometers on geostationary satellites were launched to provide high resolution measurement (Griffith et al., 1978). While geostationary IR sensors have substantial advantages in that they provide essentially continuous observations, a major limitation is that the quantity being sensed, cloud top temperature, is not directly related to precipitation (Huffman et al., 2001). Subsequently, space-borne passive microwave (PMW) radiometers evolved as a more dependable alternative (in terms of accuracy) a decade later. PMW sensors work on the principle that naturally emitted radiation in the microwave frequencies greater than 20 GHz is dictated by the composition of atmospheric hydrometeors. PMW sensors are considered more accurate under most conditions for precipitation estimation over land than their IR counterparts.

In 1997, the Tropical Rainfall Measuring Mission (TRMM), the first space-borne active microwave (AMW) precipitation radar (TRMM-PR), was launched. Although radar generally is the most accurate remote sensing technique for precipitation estimation, radar technology is costly, and TRMM-PR has limited spatial coverage (at latitudes between about 35° S and 35° N) with a sampling frequency about once per day. Therefore, the constellation of PMW sensors continue to represent a compromise between IR sensors and TRMM-PR in terms of sampling frequency, accuracy, and global coverage. GPM is therefore being planned now as a global constellation of low earth orbiting satellites (some of them existing) carrying various PMW sensors (Smith et al., 2007). It will essentially be an extension of the TRMM mission in space and time, which would provide near-global coverage of land areas, and would formally incorporate a means of combining precipitation radar with PMW sensors to maximize sampling and retrieval accuracy. The GPM Core satellite will be similar in concept to the TRMM satellite, and will house precipitation radar of improved accuracy as well as a PMW sensor (Figure 1, right panel). Through this configuration, GPM aims to provide coherent global precipitation products with temporal resolution ranging from 3 to 6 hours and spatial resolution in the range 25–100 km^2 (Smith et al., 2007; also http://gpm.gsfc.nasa.gov).

A major benefit offered by the GPM program would be the increased availability of microwave rainfall data that will be cooperatively provided from multiple platforms by several independent programs at a high temporal resolution (~3 hours). It must however be noted by the hydrologist that, the microwave overpasses yield only instantaneous rainfall estimates rather than accumulated rainfall totals that are typically used as input in hydrologic models. Caution and thoughtful preprocessing are needed before investigating the usefulness of satellite rainfall data for flood prediction (discussed in detail later). Furthermore, hydrologists need to be cognizant of the current availability of a large number of 'combined' satellite algorithms that function on the basis of both geostationary IR and PMW data. It is currently not known what role IR-based algorithms, if any, will continue to play for flood prediction during the GPM-era as the frequency of the more accurate PMW data increases many folds. Promising newer algorithms that combine the IR data more intelligently and yet manage to retain the strength of PMW algorithms should be kept in the hydrologist's shortlist of potential input data sources over ungauged basins (Joyce et al., 2004).

3 Current Knowledge Gaps on Satellite-Based Flood Prediction

3.1 The Process-Based Knowledge Gap

Understanding the current knowledge gaps on satellite based flood prediction is critical to successful application of satellite rainfall data over regions lacking access to a conventional rainfall data source. The central theme on the current knowledge gap deals with the hydrologic implications of uncertainty of the satellite rainfall estimates. This satellite estimation uncertainty manifests as a result of the mismatches in the identification of rainy and non-rainy areas by the satellite algorithm, while considerable hydrologic implication exists for this uncertainty due to the spatial scaling properties of the river basin (Wood et al., 1990). The focus of this chapter is however, mostly on the former issue (i.e., satellite rainfall uncertainty) in an independent manner, even though we recognize that the basin scaling properties would play a definitive role in dictating the optimal use of satellite rainfall data for flood prediction.

Although there are several sources of uncertainty that complicate our understanding of flood prediction accuracy (see for example, Georgakakos et al., 2004), the principal source of uncertainty is, undoubtedly, rainfall (Kavetski et al., 2006a, 2006b; Hossain et al., 2004a, 2004b; Krzyzstofowicz, 1999, 2001). In a recent study, Syed et al. (2004) corroborated this further by demonstrating that 70%-80% of the variability observed in the terrestrial hydrologic cycle is, in fact, attributed to rainfall. For the case of satellite rainfall estimation, this uncertainty can lead to unacceptably large uncertainties in runoff simulation (Nijssen and Lettenmaier, 2004). Thus, if satellite rainfall data are to be critically assessed of the opportunities they possess for river flood prediction over large ungauged basins, it is important that we first understand the error propagation that is associated with satellite-estimated rainfall.

An error propagation of satellite rainfall estimates for flood prediction applications requires the derivation of the probability distribution of simulated stream flow involving the following three components: (1) a probabilistically formulated satellite rainfall model that can simulate realistic and equi-probable random traces of satellite-like rainfall estimates; (2) a deterministic or probabilistic hydrologic model for the rainfall-runoff transformation to floods; and (3) a Monte Carlo (MC) framework linking (1) and (2). The fully random MC sampling can be currently considered the most preferred method for such uncertainty analysis due to ever-increasing computing power (Hossain et al., 2004c). Other reasons for the widespread preference of MC techniques are their lack of restrictive assumptions and completeness in sampling the error structure of the random variables (Beven and Freer, 2001; Kremer, 1983).

However, the traditional MC approach to modeling stream-flow error propagation exhibits limited physical insight of the role played by each hydrologic process control comprising the flood phenomenon. This is because the hydrologic model in a typical MC uncertainty assessment would be applied as a black-box unit for transforming the rainfall to runoff. While the more sophisticated physically-based and fully distributed hydrologic models are capable of simulating the individual water cycle components in the continuum of space and time, the problem of identifying the make up of streamflow as a function of various runoff components nevertheless persists. The derived runoff error distribution is thus consequently marginal, regardless of the type of hydrologic model use (conceptual-lumped or physically-based and distributed) because of its functional integration over the major hydrologic processes on the land surface. This distribution can not be isolated into components that can be linked directly to the individual hydrologic process controls or the nature of its physical representation (such as, infiltration, base flow, evaporation, etc.). To the best of our knowledge, there has not been any successful attempt to relate the marginal distribution of streamflow simulation error to these individual process controls. Existing literature provides little indication of a coherent agenda to explore the role of hydrologic process controls in the context of advancing satellite-based flood prediction over ungauged river basins.

But how exactly does the study of flood prediction uncertainty as a function of hydrologic process controls and satellite rainfall estimation error help in serving the greater scientific agenda of protecting mankind from the flooding hazard? For any given river basin, flood prediction needs are unique (e.g., one may be interested in stream-flow prediction at the basin outlet or distributed simulation of water levels for the entire river network). A hydrologist is faced with a wide variety of geology, soils, initial wetness, vegetation, land use and topographic characteristics that affect the relationship between rainfall and runoff in the most unique ways. This relationship consequently affects the relationship between rainfall estimation error and runoff simulation error. While detailed information on the land surface may not always be available, especially for ungauged basins, approximate characteristics such as dominant overland flow mechanism (saturation excess vs. infiltration excess), extent of evapotranspiration (low vs. high vegetation), flow regime in channels (low-mild vs. steep channel slopes) are reasonable to be known a priori. Thus, if the role played by each hydrologic process control in transforming the rainfall estimation error to stream-flow error could be understood, then, ideally, one would be better poised to wisely select a hydrologic model with "commensurate" process representation that yields "acceptable" error propagation. These would allow the hydrologist to make an informed decision on his choice for a hydrologic model for flood mitigation purposes in an ungauged basin on the basis of the quality of satellite rainfall data available to him.

3.2 The Scale-Based Knowledge Gap

Another knowledge gap that we must recognize in the context of flood prediction is the complexity of the error structure of satellite rainfall data. Unlike radar rainfall estimation, where after careful quality control and error adjustments, the residual error is associated primarily with a random component that usually has modest space-time correlation, satellite precipitation retrieval uncertainty is associated with correlated rain/no-rain detection and false alarm error characteristics as well as systematic and random rain rate error components with non-negligible spatio-temporal correlation lengths. Hossain and Anagnostou (2006a) have recently demonstrated the complex nature of this satellite error structure using a ground validation site over the

Oklahoma Mesonet region. Furthermore, different satellite rainfall algorithms would have different error characteristics, while the combined multi-sensor algorithms may be expected to have a more complex error structure depending on the type of calibration data used in the making. Nonetheless, most attempts to characterize errors in satellite precipitation retrievals to date portray the error structure using metrics that can be argued as overly simplistic, and ultimately misleading relative to the hydrologic potential of GPM. For instance, error metrics limited to 'bias' and 'random error' parameters have been used to define the minimum success criteria of GPM and other community efforts like the Pilot Evaluation of High Resolution Precipitation Products (PEHRP, Turk et al., 2006). For flood prediction, these metrics are probably not adequate, even though they may serve a very useful purpose at meteorological scales. The desire for progression to finer (spatial) scales in satellite precipitation estimation is in fact counter-balanced by increasing dimensionality of the retrieval error, which has a consequently complex effect on the propagation through land surface-atmosphere interaction simulations. This in turn has tremendous implications for the spatial and temporal scales at which hydrologic models can reasonably be implemented, or rather, the scale at which optimal data use is feasible.

As an example, consider the dynamic process of vertical soil moisture transport. The water flux in soil is governed by the cumulative effect of infiltration, runoff, gradient diffusion, gravity, and soil water extraction through roots for canopy transpiration. All these processes exhibit dynamic variability in the ranges of minutes to hours over scales of cm^2 to km^2. However, satellite precipitation algorithms of today cannot hope to resolve these resolutions. It is even doubtful if the future space-borne precipitation remote sensing can independently deliver the rainfall data at the resolution where surface hydrology is dominant, which is at considerably smaller space-time scales than the typically coarser meteorologic scale at which satellite data is produced. As a minimum, there is a need to understand the spatial resolution to which satellite products can realistically be disaggregated (see Margulis and Entekhabi, 2001 and Venugopal et al., 1999 for example) and to estimate the resulting error structure, and its interaction with hydrologic models which produce flood forecasts. The scale incongruity between meteorological process data and its hydrologic application represents a competing trade off between lowering the satellite retrieval error versus modeling land-vegetation-atmosphere processes at the finest scale possible. While much remains to be done to define these trade-offs towards optimal use of satellite rainfall data in hydrologic models, it may well not be possible to implement GPM products at scales as fine as those cited above (e.g., 5 km), and that the 25-100 km resolutions suggested by Smith et al. (2007) may perhaps be more realistic and reliable for the hydrologist for flood prediction.

A major problem encountered in application of satellite rainfall data is that the frequency of complex mis-matches (with the ground-truth) increases as the satellite rainfall data is progressively reduced in scale (as alluded earlier

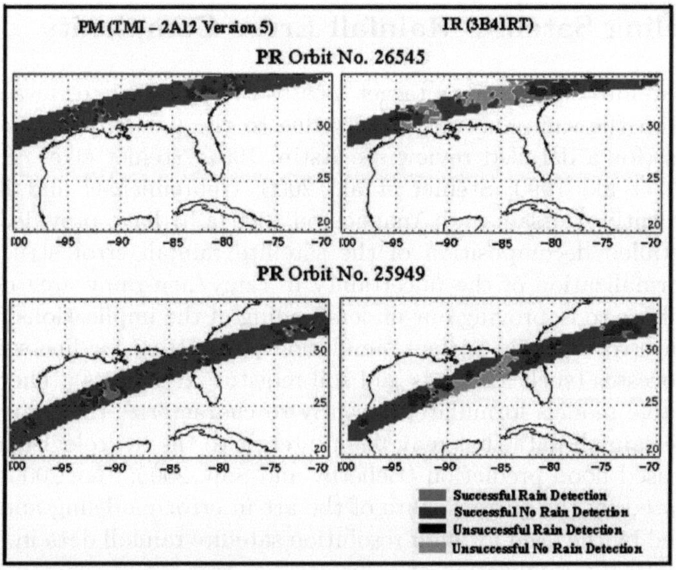

Fig. 2. Successful and unsuccessful rain and no-rain detection by MW and IR sensors referenced with TRMM-PR observations.

in Section 3.1). We can demonstrate this phenomenon through an example on the detection performance of two types of satellite rainfall types. Figure 2 demonstrates typical detection capabilities for rain and no-rain for two different sensors (PMW – left panel; IR-right panel) using the most accurate space-borne rainfall data derived from the TRMM Precipitation Radar (PR) as 'ground-truth' (data product name 2A25). The presence of definitive spatial structures of detection of rain and no-rain as a function of sensor-type is clearly evident. This detection capability is also known to be strongly influenced by scale and rainfall rates (Hossain and Anagnostou, 2006a).

The success in resolving the scale incongruity to a level practically feasible for flood prediction will therefore rest on the feedback between hydrologists and meteorologists (the typical algorithm and data producers). Even though the efforts at addressing hydrologic prediction uncertainty (Beven and Binley, 1992) are probably as mature as the efforts to characterize uncertainty of remote sensing of rainfall (North and Nakamoto, 1989), both efforts have evolved independently. This lack of feedback can be attributed to the absence of proper metrics and frameworks that are interpretable by both end-user hydrologists and producer meteorologists. Two satellite rainfall algorithms with similar bias and root mean squared error (RMSE) can have much different error propagation properties when used in hydrologic models (Lee and Anagnostou, 2004). Thus hydrologists today are therefore left with inadequate metrics to identify optimal data use and thereby communicate to the data producers on the desired minimum criteria for a satellite mission to be effective in flood prediction at pertinent scales and geographic location.

4 Modeling Satellite Rainfall Error Complexity

Current satellite error models target mostly PMW sensor retrievals focusing primarily on the sampling uncertainty due to the low frequency of satellite overpasses (for a detailed review see Astin, 1997; Steiner et al., 1996; Bell, 1987; Bell et al., 1990; Steiner et al., 2003; Gebremichael and Krajewski, 2004). Recently, Hossain and Anagnostou (2004a,b) have provided evidence that a detailed decomposition of the satellite rainfall error structure with explicit formalization of the uncertainty in rainy/non-rainy area delineation can contribute to improving our understanding of the implications of satellite estimation error on land surface simulation parameters for fine-scale hydrological processes (such as floods and soil moisture dynamics). There are also other notable models formulated recently to characterize the error structure of satellite rainfall data that may be of interest to the hydrologist to advance satellite based flood prediction (Bellerby and Sun, 2005; Teo, 2006).

Motivated by the current state of the art in error modeling and the challenges faced by the need for high-resolution satellite rainfall data in hydrology, a mathematical formalization of a space-time error model, named *SREM2D*, was recently developed by Hossain and Anagnostou (2006a). *SREM2D* had the following design objectives in mind during its conceptualization: (1) It should function as a filter wherein the hydrological implications of fine-scale components of the satellite precipitation error structure can be explicitly determined by coupling it with a hydrological/land surface model; Thus, *SREM2D*-based experiments should provide the much needed focus to meteorologists for the development of next-generation of satellite rainfall products with enhanced societal applications; These experiments should also help hydrologist identify the optimality criterion for using a given satellite rainfall dataset in a hydrologic model; (2) It should be modular in design with the capability to allow uncertainty assessment of *any* satellite rainfall algorithm; (3) It should be conceptualized in an algorithmic fashion so that it is easy to code numerically by a user wishing to make use of the model for his/her own scientific agenda. *SREM2D* uses as input "reference" rain fields of higher accuracy and resolution representing the "true" surface rainfall process, and stochastic space-time formulations to characterize the multi-dimensional error structure of satellite retrieval. The algorithmic approach of *SREM2D* is aimed at generating realistic ensembles of satellite rain fields from the most definitive "reference" rain fields that would preserve the estimation error characteristics at various scales of aggregation. By propagating the simulated ensembles in a hydrologic model, *SREM2D* would therefore allow the understanding of the implications of satellite rainfall error structure and scale complexity on streamflow simulation.

The major dimensions of error structure in satellite estimation modeled by *SREM2D* are (1) the joint probability of successful delineation of rainy and non-rainy areas accounting for a spatial structure; (2) the temporal dynamics of the conditional rainfall estimation bias (rain $>$ 0 unit); and (3) the spatial structure of the conditional (rain > 0 unit) random deviation. The spatial structure in *SREM2D* is modeled as spatially correlated Gaussian random fields while the temporal pattern of the systematic deviation is modeled using a lag-one autoregressive process. The spatial structures for rain and no-rain joint detection probabilities are modeled using Bernoulli trials of the uniform distribution with a correlated structure. This correlation structure is generated from Gaussian random fields transformed to the uniform distribution random variables via an error function transformation. There are nine error parameters in total. Complete details on *SREM2D* can be found in Hossain and Anagnostou (2006a).

5 Current Progress on Closing the Knowledge Gap

5.1 On Scale-Based Knowledge Gap

Comparison of *SREM2D*-simulated satellite rainfall with actual satellite rainfall data produced by NASA (IR-3B41RT; Huffman et al., 2007) has shown that a complex and multi-dimensional error modeling technique (such as *SREM2D*) can preserve the estimation error characteristics across scales with marginal deviations. Upon comparison with less complex and commonly applied error modeling strategies, it was also shown that these (simpler) approaches typically underestimated sensor retrieval error standard deviation by more than 100% upon aggregation, which, for *SREM2D*, was found to be below 30% (Hossain and Anagnostou, 2006a). More recent studies have further demonstrated that understanding of the hydrologic implications of satellite-rainfall data overland can be significantly improved through the use of *SREM2D* in a hydrologic error propagation framework. This is a promising finding as it would allow a more reliable investigation of the optimality criterion for using satellite rainfall data in hydrologic models. Two aspects were examined in detail: (1) soil moisture dynamics and (2) ensemble rainfall data generation. For understanding the impact of satellite rainfall uncertainty on soil moisture dynamics, the Common Land Model (CLM; Dai et al., 2003) was coupled with *SREM2D* to propagate ensembles of simulated satellite rain fields for the prediction of soil moisture at 5 cm depth region. It was observed that *SREM2D* captured the spatiotemporal characteristics of soil moisture uncertainty with higher consistency than a simpler bi-dimensional error modeling strategy (Figure 3, upper panels; Hossain and Anagnostou, 2005b). In a subsequent follow-up study, further insights were revealed from the pursuit of the scientific query: *Can a multidimensional satellite rainfall error model perform realistic ensemble generation of satellite rainfall data of improved accuracy for a satellite retrieval technique?* Using as reference, ground radar (WSR-88D) rainfall fields, the scale-dependent multidimensional error structure for satellite rainfall algorithms was determined. Next, by reversing the definition of reference and corrupted rain fields produced by *SREM2D*, the inverse multidimensional error structure of WSR-88D rainfall fields with respect

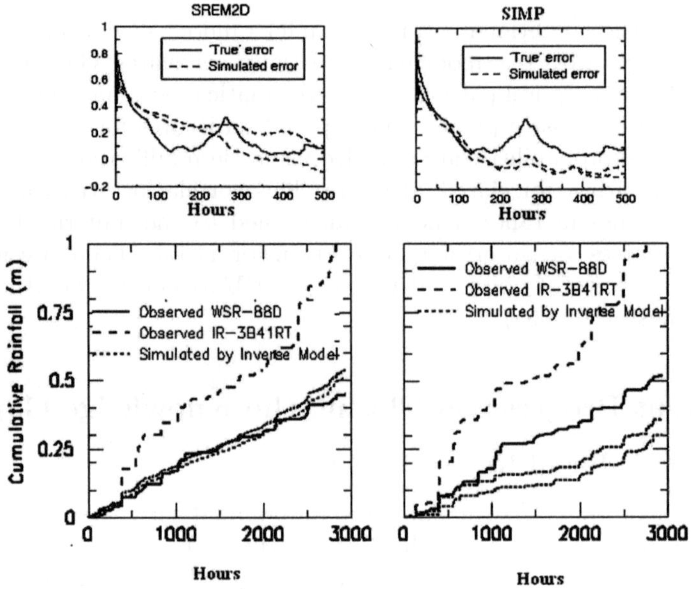

Fig. 3. Hydrologic implications of using a multidimensional satellite rainfall error modeling strategy such as *SREM2D*.

to the satellite rainfall data was identified. *SREM2D* was then run in the inverse mode to generate reference-like realizations of rainfall. The accuracy of inverse-SREM2D rainfall ensemble was observed to be consistently higher than the simpler inverse error-modeling scheme for the IR-3B41RT product (Figure 3, lower panels).

Because most attempts to characterize errors in satellite precipitation retrievals to date portray the error structure of satellite rainfall estimates using metrics that are overly simplistic, and ultimately misleading relative to the true hydrologic potential of satellite rainfall data, a complex and multidimensional error modeling strategy that is compatible with the dynamic nature of land surface hydrologic processes is needed to advance optimal data use in hydrologic models.

Upper Panels

Temporal correlogram of error field for near surface (5 cm) soil moisture simulated by CLM driven by simulated satellite rainfall data based on two error modeling schemes—*SREM2D* (left panel) and SIMP (right panel). The solid line represents the true soil moisture error dynamics on the basis of actual IR-3B41RT data. The dashed line represents the range of soil moisture error dynamics based on simulation by error model. SIMP represents the commonly used error modeling strategy in literature based on simple error statistics

(From Hossain and Anagnostou, 2005b; *Reprinted with kind permission from American Geophysical Union*).

Lower Panels

Ensemble envelopes (i.e. uncertainty range) of satellite-retrieved cumulative hyetographs (dotted lines) for two error-modeling schemes—*SREM2D* (left panel) and SIMP (right panel). Solid line represents "true" rainfall cumulative hyetograph from WSR-88D estimates, while the dashed line is the rainfall cumulative hyetograph from actual IR-3B41RT data. (From Hossain and Anagnostou, 2006b; *Reprinted with kind permission from the Institution of Electrical and Electronic Engineers*).

5.2 On Process-Based Knowledge Gap

Use of a Modular Hydrologic Modeling Platform

In order to understand the implication of satellite rainfall error on hydrologic processes, we recently developed an open-book watershed model (Figure 4). The design was modular wherein specific hydrologic processes could be conveniently altered or added to make a process-based understanding of satellite rainfall error propagation (as discussed in Section 3.1). A square-grid volume domain was used where the individual processes of overland flow and infiltration to the subsurface were linked to simulate the response of the unsaturated zone to rainfall (Figure 1). In the open-book model, the generated surface and subsurface runoff were calculated at each time-step from knowledge of the time-varying infiltration (or recharge to the soil) and by keeping track of the soil water storage for each grid volume at every timestep. The overland flow was then routed along the direction of steepest gradient for each grid surface until it laterally drains into the main channel. The streamflow was modeled as

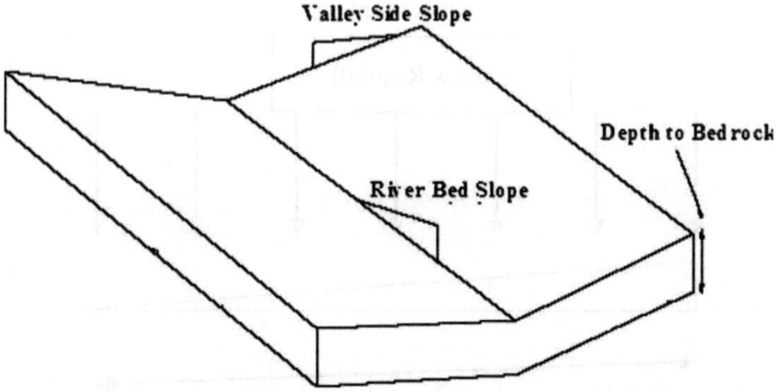

Fig. 4. Geometric representation of the open-book watershed topography. Here, the depth to bedrock basically refers to the assumed depth of the effective soil column.

a 1-D kinematic flow using Manning's equation. Evapo-transpiration and 2-D saturated zone flow were assumed insignificant in our conceptualizations as our goal was to focus primarily on streamflow simulation at the timescales of flooding where the surface and the unsaturated zones are considered hydrologically the most dynamic regions. A point to note herein is that the 'Depth to bedrock' shown in Figure 4 is basically the depth of the effective soil column. Complete details on the open-book watershed model can be found in Katiyar and Hossain (2007).

The Hydrologic Process Conceptualizations

To understand the role played by specific hydrologic process control conceptualization, three types of rainfall-runoff conceptualizations were considered for computing excess rainfall over a grid volume (see Figure 5). These conceptualizations were: (1) A simple statistical parameterization to compute excess rainfall; (2) A linear storage-discharge conceptualization for surface and subsurface runoff generation; (3) A non-linear storage-discharge conceptualization for surface and subsurface runoff generation. The overland and river flow components of the model remained the same. These process conceptualizations employed basically a mass-balance approach and are presented briefly as follows.

For the statistical model, the precipitation $p(t)$ was partitioned into infiltration to the soil water store as $\boldsymbol{a}p(t)$, and surface runoff (quickflow/overland flow) as $(1 - \boldsymbol{a})p(t)$. The subsurface flow draining from the grid volume's soil water store is assumed insignificant (at flooding timescales) and the soil water storage is updated at each time-step on the basis of the recharge only. When it equaled the maximum storage capacity of S_b (computed as $D\phi$; D is depth of effective soil column and ϕ is porosity), all precipitation was consequently transformed as surface runoff with no recharge.

For the linear storage-discharge conceptualization, the following water balance equation was used for each grid volume,

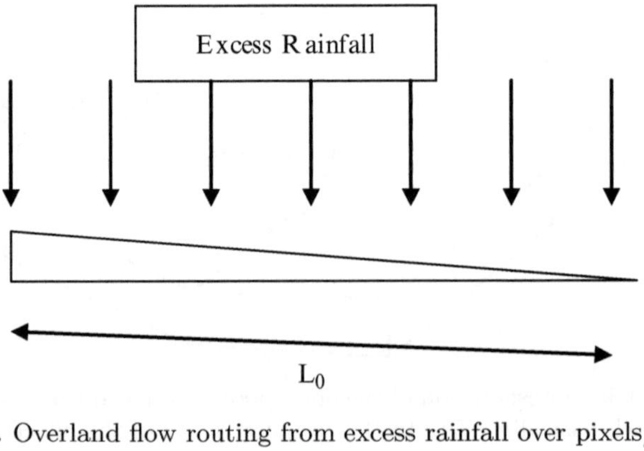

Fig. 5. Overland flow routing from excess rainfall over pixels/zones.

$$\frac{ds(t)}{dt} = p(t) - q_{se}(t) - q_{ss}(t) \tag{1}$$

where, $S(t)$ is the soil water storage, $p(t)$ is the precipitation, $q_{se}(t)$ is the overland saturation-excess flow and $q_{ss}(t)$ is the sub-surface flow at time t. The $q_{ss}(t)$ and $q_{se}(t)$ were computed as follows,

$$q_{ss} = \frac{S(t) - S_f}{t_c} \quad \text{if } S(t) > S_f \tag{2a}$$

$$q_{ss} = 0 \quad \text{if } S(t) < S_f \tag{2b}$$

where, S_f is the soil moisture storage at field capacity (defined by the soil type) and t_c is the grid response time to subsurface flow. t_c is approximated from Darcy's law assuming a triangular groundwater aquifer and hydraulic gradient approximated by ground slope.

$$t_c = \frac{L\phi}{2K_s \tan\beta} \tag{2c}$$

Herein, L is the grid size, K_s the saturated hydraulic conductivity and β is the grid slope. The sub-surface flow draining out from each grid volume is not routed within the soil medium for the same reason that it would comprise an insignificant component of the total flood volume. However, this model conceptualization represented a complexity level higher than the previous statistical parameterization because of the use of mass balance equation and physically-based watershed parameters to identify the saturation excess runoff. The overland saturation excess flow $q_{se}(t)$ was computed as follows,

$$q_{se} = \frac{S(t) - S_b}{\Delta t} \quad \text{if } S(t) > S_b \tag{3a}$$

$$q_{se} = 0 \quad \text{if } S(t) < S_b \tag{3b}$$

where S_b is the soil's storage capacity computed as $D\phi$ (D is effective soil column depth and ϕ is porosity).

Finally, for the non-linear storage-discharge conceptualization, the subsurface runoff in the linear model was reformulated as follows,

$$q_{ss} = \left[\frac{S(t) - S_f}{a}\right]^{\frac{1}{b}} \quad \text{if} S(t) > Sf \tag{4a}$$

$$q_{ss} = 0 \quad \text{if } S(t) < S_f \tag{4b}$$

Here, parameter b defines the degree of non-linearity in the storage-discharge relationship, while a replaces t_c in Equation (2a). Figure 6 summarizes all three process conceptualizations to showcase the gradual increase in model complexity.

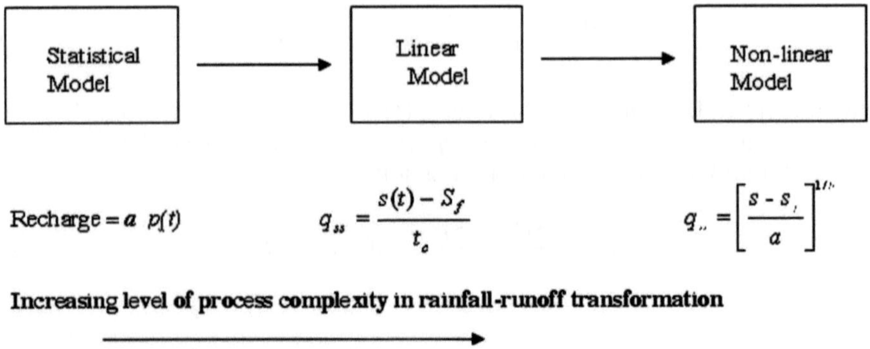

Fig. 6. The three rainfall-runoff process conceptualizations in the order of increasing complexity from left to right. The process difference is shown along with the model name.

The Error Propagation Framework

SREM2D was coupled to the open-book models to understand the response of satellite rainfall error to spatial scaling on river flow prediction uncertainty. We used, as our reference, quality-controlled ground radar (WSR-88D) rainfall data over the Oklahoma Mesonet region. Satellite rainfall error parameters were derived for satellite rainfall algorithm (3B41RT) that is produced by NASA (Huffman et al., 2007) on a real-time basis. *SREM2D* was then used to corrupt radar rainfall fields in a space-time framework to simulate satellite-like rain estimates. The satellite rainfall error propagation in streamflow prediction was assessed in a MC framework for the three model types across two spatial scales of aggregations –0.25 degree and 0.50 degree. The 15 MC realizations of *SREM2D* generated rainfields that were propagated through each open-book model conceptualization yielded corresponding uncertainty limits in streamflow simulation.

Two contrasting issues were considered in the error propagation. If either the uncertainty limits were predicted too narrowly or the whole ensemble envelope is biased (i.e., the reference streamflow is consistently outside the prediction error bounds), then a comparison with in-situ/reference measurements would suggest that the combined rainfall-model complexity structure was invalid for the satellite rainfall error. If, on the other hand they were predicted too widely, then it could be concluded that the hydrologic modeling structure had little predictive capability. The dichotomous nature of '*structural validity*' and '*predictive ability*' was quantified by the *Exceedance Probability* (*EP* in Equation (5)) and *Uncertainty Ratio* (*UR* in Equation (6)), respectively, as follows:

$$EP = \frac{\text{Number of times reference streamflow exceeds the uncertainty limits}}{\text{Total number of timesteps}} \tag{5}$$

Table 1. *UR* and *EP* values as a function of spatial scale and process conceptualization.

0.25 degree		0.50 degree	
UR	EP	UR	EP
Statistical Model			
0.588	0.607	0.670	0.3064
% Change upon aggregation to 0.5 degree		+14.0%	−40.0%
Linear Model			
0.492	0.514	0.594	0.450
% Change upon aggregation to 0.5 degree		+20%	−12.5%
Non-Linear Model			
0.561	0.557	0.668	0.476
% Change upon aggregation to 0.5 degree		+19%	−14.5%

$$UR = \frac{\text{Uncertainty in runoff volume simulation (beween uncertainty limits)}}{\text{Observed Runoff Volume}} \tag{6}$$

Table 1 summarizes the findings of the error propagation experiment as a function of scale and hydrologic process conceptualizations. The global picture that emerges from this table can be summarized as follows:

(1) statistical parameterization for excess rainfall results in increased sensitivity of satellite rainfall error to streamflow prediction uncertainty; this sensitivity, however, responds favorably to scaling towards improving the model's structural validity at larger scales of aggregation;
(2) inclusion of a linear/non-linear reservoir for subsurface flow representation visibly smoothens the hydrologic simulations and reduces the runoff simulation uncertainty;
(3) insignificant beneficial impact is observed through the inclusion of non-linearity in the storage-discharge relationship and it may so be that the scale of application is already so large that satellite rainfall error is insensitive to any further increase in process complexity;
(4) there is strong indication that hydrologic process complexity plays a definitive role in accurately capturing streamflow variability on the basis of model driven by downscaled satellite input data.

While the findings represented a useful first step, our study has limitations like any other investigation. Hence, it is important to extend the investigation involving a wider range of research objectives and more complex hydrologic process representation. Examples of extension could be: (1) repeat the investigation using real-world watersheds in a range of climatic conditions, distributed hydrologic models and elevation data and thereby understand the utility of the open-book approach as a physically consistent proxy for investigating optimality criteria; (2) increase the complexity of the hydrologic processes through more physically-based process equations (i.e., Richards equation or Green and Ampt for infiltration; energy-balance method for representation of evapo-transpiration etc.); and finally, (3) explore scaling behavior at finer space-time resolutions (<0.25 degree and <1 hourly). Currently the Land Information System of NASA (LIS) provides the hydrologic state of the land at 0.5 hour $1 \times 1\,\mathrm{km}^2$ resolution using satellite datasets that are subsequently ingested for numerous societal applications such as weather prediction, agricultural planning, army operations etc (Kumar et al., 2006). We hope that extension of our work along these directions can consequently help us achieve a firmer understanding of the optimality criteria for use of remotely-sensed rainfall data from space-borne platforms in hydrologic models.

6 Concluding Remarks

For advancing the use of satellite rainfall data for flood prediction, there can be two major issues related to rainfall data uncertainty that hydrologists need to recognize in the context of flood prediction– i) the role played by hydrologic process controls; and ii) the role played by scale. We have highlighted the progress made by us on the understanding of these two issues. Much work remains to be done towards a more complete understanding on optimal use of satellite rainfall data in hydrologic models. It is however, equally important to initiate the work in anticipation of a successful leveraging of GPM. We have argued in this chapter, as has been argued previously by others, that unless there is a shift in paradigm, the conventional assessment frameworks and metrics for estimation of rainfall from satellite sensors will probably remain inadequate for hydrologic purposes such as flood prediction. We also argued that greater emphasis must be placed on development of hydrologically relevant precipitation estimation algorithms, and that this will require involvement of a broader cross-section of the hydrologic community. We therefore hope that identification of these key issues, as discussed in this chapter, will usher a new era for hydrologists working on optimal use of satellite rainfall data in anticipation of GPM.

We would like to close this chapter with a candid discussion of the limitations and disclaimers associated with our study that readers should be aware of. For example, while we have predominantly focused on floods, the choice of appropriate error metrics would most probably be dictated by the flood type (high/extreme floods versus low/frequent floods). Furthermore, the hydrologic implication of satellite rainfall error would be strongly influenced by the hydrologic variable (or predictand) in question. Again, this chapter focused on floods, while soil moisture, which plays a critical role in partitioning of rainfall into runoff, would not have the same implication as streamflow. The reader is referred to the work of Hossain and Anagnostou (2005a, 2005b) where a detailed investigation has been carried out for soil moisture.

Acknowledgements

The authors gratefully acknowledge the comments received from two anonymous reviewers. The first author was supported by the Ivanhoe Fellowship and a research initiation grant from the Tennessee Technological University. Additional support from the Center for Management, Utilization and Protection of Water Resources of Tennessee Technological University and NASA's Precipitation Measurement Mission are also acknowledged. Sections of the work reported in this chapter are based on recent research documents published by the authors and are referenced here only as a means of ensuring timely dissemination of scholarly and technical work on Hydrologic data fusion.

References

Anagnostou, E.N. (2004). Overview of Overland Satellite Rainfall Estimation for Hydro-Meteorological Applications. *Surveys in Geophysics*, 25(5–6), pp. 511–537.

Astin, I. (1997). A survey of studies into errors in large scale space-time averages of rainfall, cloud cover, sea surface processes and the earth's radiation budget as derived from low orbit satellite instruments because of their incomplete temporal and spatial coverage. *Surveys in Geophysics*, 18, pp. 385–403.

Bell, T.L. (1987). A space-time stochastic model of rainfall for satellite remote-sensing studies. *Journal of Geophysical Research*, 92D, pp. 9631–9643.

Bell, T.L., Abdullah, A., Martin, R.L., and North, G.R. (1990). Sampling errors for satellite-derived tropical rainfall—Monte Carlo study using a space-time stochastic model. *Journal of Geophysical Research*, 95D, pp. 2195–2205.

Bellerby, T., and Sun, J. (2005). Probabilistic and ensemble representations of the uncertainty in IR/Microwave precipitation product. *Journal of Hydrometeorology*, 6, pp. 1032–1044.

Beven, K.J., and Freer, J. (2000). Equifinality, data assimilation, and uncertainty estimation in mechanistic modeling of complex environmental systems using the GLUE methodology. *Journal of Hydrology* 249, pp. 11–29.

Beven, K.J., Binley, AM. (1992). The future of distributed models: model calibration and uncertainty prediction. *Hydrological Processes*, 6, pp. 279–298.

Gebremichael, M., and Krajewski, W.F. (2004). Characterization of the temporal sampling error in space-time-averaged rainfall estimates from satellites.*Journal of Geophysical Research* 109(D11), (doi: D11110 0.1029/2004JD004509).

Georgakakos, K.P., Seo, D-J., Gupta, H., Schaake, J.C., and Butts, M. (2004). Towards the characterization of stream-flow simulation uncertainty through multimodel ensembles. *Journal of Hydrology*, 298, pp. 223–241.

Griffith, C.G., Woodley, W.L., and Grube, P.G. (1978). Rain estimation from geosynchronous satellite imagery-visible and infrared studies. *Monthly Weather Review* 106, pp. 1153–1171.

Hossain, F. (2006). Towards formulation of a fully space-borne system for early warning of floods: Can cost-effectiveness outweigh flood prediction uncertainty? *Natural Hazards*, 37(3), pp. 263–276 (doi :10.1007/s11069-005-4645-0).

Hossain, F., and Anagnostou, E.N. (2006a). A two-dimensional satellite rainfall error model. *IEEE Transactions Geosciences and Remote Sensing*, 44(6), pp. 1511–1522 (doi: 10.1109/TGRS.2005.863866).

Hossain, F., and Anagnostou, E.N. (2006b). Assessment of a multi-dimensional satellite rainfall error model for ensemble generation of satellite rainfall data. *IEEE Geosciences and Remote Sensing Letters*, 3(3), pp. 419–423.

Hossain, F., and Lettenmaier, D.P. (2006). Flood Prediction in the Future: Recognizing Hydrologic Issues in Anticipation of the Global Precipitation Measurement Mission. *Water Resources Research*, 44, (doi:10.1029/2006WR005202) .

Hossain, F., and Katiyar, N. (2006). Improving flood forecasting in international river basins. *EOS Transactions (AGU)* 87(5), pp. 49–50.

Hossain, F., and Anagnostou, E.N. (2005a). Numerical investigation of the impact of uncertainties in satellite rainfall estimation and land surface parameters for simulation of soil moisture. *Advances in Water Resources,* 28(12), pp. 1336–1350 (doi:10.1016/j.advwatres.2005.03.013).

Hossain, F. and Anagnostou, E.N. (2005b). Using a multi-dimensional satellite rainfall error model to characterize uncertainty in soil moisture fields simulated by an offline land surface model *Geophysical Research Letters*, 32(L15402) (doi:10.1029/2005GL023122).

Hossain, F., Anagnostou, E.N., and Dinku, T. (2004a). Sensitivity analyses of satellite rainfall retrieval and sampling error on flood prediction uncertainty. *IEEE Transactions on Geosciences and Remote Sensing*, 42(1).

Hossain, F., Anagnostou, E.N., Dinku, T., and Borga, M. (2004b). Hydrological model sensitivity to parameter and radar rainfall estimation uncertainty. *Hydrological Processes* 18 (doi:10.1002/hyp.5659).

Hossain, F., Anagnostou, E.N., and Lee K-H. (2004c). A non-linear and stochastic response surface method for Bayesian estimation of uncertainty in soil moisture simulation from a land surface model. *INVITED PAPER—Special Issue in Non-Linear Processes in Geophysics*, 11, pp. 1–15 (SREF-ID: 1607-7946/npg/2004-11-1).

Huffman, G.J., Adler, R.F. Morrissey, M.M. and others. (2001). Global precipitation at one-degree daily resolution from multisatellite observations. *Journal of Hydrometeorology*, 2, pp 36–50.

Huffman, G.J., Adler, R.F., Bolvin, D.T., Gu, G., Nelkin, E.J., Bowman, K.P., Hong, Y., Stocker, E.F., and Wolff, D.B. (2007). The TRMM multi-satellite precipitation analysis: Quasi-global, multi-year, combined sensor precipitation estimates at fine scales, *Journal of Hydrometeorology*, 8, pp. 28–55.

Joyce, R.J. and Janowiak, J.E., Arkin, P.E., and Xie, P. (2004). CMORPH: A method that produces global precipitation estimates from passive microwave and infrared data at high spatial and temporal resolution. *Journal of Hydrometeorology*, 5(3), pp. 487–503.

Katiyar, N., and Hossain, F. (2007). An open-book watershed model for prototyping space-borne flood monitoring systems in international river basins. *Environmental Modeling and Software*, 22(12), pp. 1720–1731 (doi:10.1016/j.envsoft.2006.12.005).

Kavetski, D., Kuczera, G., and Franks, S.W. (2006a). Bayesian analysis of input uncertainty in hydrological modeling: 2 Application. *Water Resources Research*, 42(W03408) (doi:10.1029/2005WR004376).

Kavetski, D., Kuczera, G., and Franks, S.W. (2006b). Bayesian analysis of input uncertainty in hydrological modeling: 1 Theory. *Water Resources Research*, 42 (W03407) (doi:10.1029/2005WR004368).

Kremer, J.N. (1983). Ecological implications of parameter uncertainty in stochastic simulation. *Ecological Modeling* 18:187–207.

Krzyzstofowicz, R. (1999). Bayesian theory of probabilistic forecasting via deterministic hydrological model. *Water Resources Research* 35(9), pp. 2739–2750.

Krzyzstofowicz, R. (2001). The case for probabilistic forecasting in hydrology. Journal of Hydrology 249, pp. 2–9.

Kumar, S.V., Peters-Lidard, C.D., Tian, Y., Geiger, J., Houser, P.R., Olden, R., Lighty, L., Eastman, J.L., Dirmeyer, P., Doty, P., Adams, P.J., Wood, E.F., and Sheffield, J. (2006). LIS – An Interoperable Framework for High Resolution Land Surface Modeling. *Environmental Modeling and Software* 21(10), pp. 1402–1415. (doi. 10.1016/j.envsoft.2005.07.004).

Lee, K.H., and Anagnostou, E.N. (2004). Investigation of the nonlinear hydrologic response to precipitation forcing in physically based land surface modeling. *Canadian Journal of Remote Sensing* 30(5), pp. 706–716.

Margulis, S., and Entekhabi, D. (2001). Temporal disaggregation of satellite-derived monthly precipitation estimates and resulting propagation of error in partitioning of water at the land surface. *Hydrology and Earth System Sciences,* 5(1), pp. 27–38.

Nijssen, B., and Lettenmaier, D.P. (2004). Effect of precipitation sampling error on simulated hydrological fluxes and states: Anticipating the Global Precipitation Measurement satellites. *Journal of Geophysical Research,* 109(D02103).

North, G.R., and Nakamoto, S. (1989). Formalism for comparing rain estimation designs. *Journal of Atmospheric and Oceanic Technology,* 6, pp. 985–992.

Shiklomanov, A.I., Lammers, R.B., and Vörösmarty, C.J. (2002). Widespread decline in hydrological monitoring threatens pan-arctic research. *EOS Transactions* 83(2), pp. 16–17.

Smith E., Asrar, G., Furuhama, Y., Ginati, A., Kummerow, C., Levizzani, V., Mugnai, A., Nakamura, K., Adler, R. Casse, V., Cleave, M., Debois, M., Durning, J., Entin, J., Houser, P., Iguchi, T., Kakar, R., Kaye, J., Kojima, M., Lettenmaier, D.P., Luther, M., Mehta, A., Morel, P., Nakazawa, T., Neeck, S., Okamoto, K., Oki, R., Raju, G., Shepherd, M., Stocker, E., Testud, J., and Wood, E.F. (2007). The international global precipitation measurement (GPM) program and mission: An overview. In *Measuring Precipitation from Space: EURAINSAT and the Future,* (Eds.) V. Levizzani and F.J. Turk, Kluwer Academic Publishers (In press; copy available at http://gpm.gsfc.nasa.gov).

Steiner, M. (1996). Uncertainty of estimates of monthly areal rainfall for temporally sparse remote observations. *Water Resources Research,* 32:373–388.

Steiner, M., Bell, T.L., Zhang, Y., and Wood, E.F. (2003). Comparison of two methods for estimating the sampling-related uncertainty of satellite rainfall averages based on a large radar dataset. *Journal of Climate,* 16:3759–3778.

Stokstad, E. (1999). Scarcity of rain, stream gages threatens forecasts. *Science* 285, pp. 1199.

Syed, T.H., Lakshmi, V., Paleologos, E., Lohmann, D., Mitchell, K., and Famiglietti, J. 2004. Analysis of process controls in land surface hydrological cycle over the continental United States. *Journal of Geophysical Research* 109(D22105), doi: 10.1029/2004JD004640.

Teo, C.K. (2006). Satellite-based rainfall estimation and its application to crop yield modeling. *PhD Thesis (unpublished),* Department of Meteorology, University of Reading.

Turk, J., Bauer, P., and Ebert, E.E., and Arkin, P.A. (2005). Satellite-Derived Precipitation Verification Activities within the International Precipitation Working Group (IPWG), 14th *Conference on Satellite Meteorology and Oceanography,* P2.15.

Venugopal, V., Foufoula-Georgio, E., and Sapozhinikhov, V. (1999). A space-time downscaling model for rainfall. *Journal of Geophysical Research-Atmospheres,* 104(D16), pp. 19705–19721.

Integrated Methods for Urban Groundwater Management Considering Subsurface Heterogeneity

Jannis Epting[1], Peter Huggenberger[1], Christian Regli[2], Natalie Spoljaric[1], and Ralph Kirchhofer[1]

[1]Department of Environmental Sciences, Applied and Environmental Geology, University of Basel
[2]GEOTEST AG

Summary. Open space in urban areas is very rare and new infrastructure is increasingly constructed in the subsurface. These constructions may temporarily affect urban groundwater systems during construction and permanently after completion. As regards these impacts together with ancient contaminated industrial sites, particular focus was placed on determining the data required to understand changes affecting groundwater flow and transport. The extended knowledge of groundwater flow regimes could lead to reducing and minimizing, as far as possible, the negative impacts throughout the construction phases, and to developing sustainable groundwater use and management tools.

The consideration of subsurface heterogeneity is often based on pumping tests, leading to a characteristically large-scale zoning of aquifer parameters. This study compares groundwater modeling results from integrating large-scale zoning of aquifer parameters on the one hand, and a sedimentary structure-based heterogeneous description of the aquifer properties on the other.

This approach was applied to an ongoing subsurface highway construction northwest of the city of Basel, Switzerland – an area formerly contaminated by industrial activities. Today, urban groundwater resources are extensively used by industry. An integrated multidisciplinary approach was chosen to predict, mitigate or prevent environmental problems, as well as to ensure groundwater supply throughout construction. It includes integration of geological and hydrological data and results into a groundwater management system comprising: (1) extensive groundwater monitoring; (2) development of a database application facilitating lithofacies-based interpretation of drill-core data; (3) geostatistical analyses of the aquifer's heterogeneity and simulations of hydraulic parameter distributions as well as; (4) regional and local high-resolution groundwater modeling. The combination of techniques presented exemplifies the fusion of quantitative and qualitative geological and hydrological information of different quality.

J. Epting et al.: *Integrated Methods for Urban Groundwater Management Considering Subsurface Heterogeneity*, Studies in Computational Intelligence (SCI) **79**, 183–218 (2008)
www.springerlink.com

1 Introduction

Urban areas are characterized by an increasing number of constructions below the surface or even below the groundwater table, resulting in significant changes in groundwater quality and dynamics of both local and regional groundwater flow regimes. This includes a reduction of cross-sectional groundwater flow and aquifer-storage capacities. To develop concepts and methods for sustainable groundwater use in urban areas, environmental impact assessments not only have to include above-ground impairments, such as ground motions with effects on existing buildings and infrastructures, as well as noise exposure and air pollution, but also the negative impacts on groundwater flow regimes. The term groundwater flow regime comprises groundwater flow paths, velocities and budgets for a defined region in a temporal context. Among the various other possible sources of groundwater pollution observed in urban environments, subsurface construction may be a source of interference of a previously balanced urban groundwater flow regime.

Numerous urban areas in Central Europe and North America are located in flood plains of rivers canalized last century. Coarse gravelly sediments of braided rivers form one particular, frequently occurring environment within these valley fills. In most practical engineering studies, the subsurface is still represented as homogeneous or at least one consisting of a set of homogeneous layers. Groundwater flow and solute transport processes within these coarse, permeable sediments are strongly influenced by subsurface heterogeneities and require detailed knowledge of aquifer properties, such as hydraulic conductivity, porosity and dispersivity, together with their spatial distribution. This heterogeneity originates from sediment sorting processes in a dynamic environment of aggradational and erosional processes typical of braided river deposits (Huggenberger and Regli, 2006). Hydraulic conductivity variations over several magnitudes are of key importance for groundwater flow and solute migration (Rehfeldt et al., 1993; Adams and Gelhar, 1992; Gelhar, 1986). Since continuous 3D information on hydraulic properties cannot be obtained in fluvial sediments, different methods have been developed to map aquifer properties. Koltermann and Gorelick (1996) distinguish three main types of methods: (1) ***Structure-imitating methods*** using any combinations of Gaussian and non-Gaussian statistically and geometrically-based relationships to match observed sedimentary patterns. (2) ***Process-imitating methods*** consisting of aquifer calibration techniques, which solve governing equations of fluid flow and transport, as well as geological process models combining mass and momentum conservation principles with sediment transport equations. (3) ***Descriptive methods*** using different field methods to translate the resulting geological sedimentary structure models into hydrofacies models with characteristic aquifer properties. All these methods have already been applied in coarse glacio-fluvial gravel deposits typical of braided river environments (e.g. Bridge and Lunt, 2006; Nowak and Cirpka, 2006; Rubin et al., 2006; Teles et al., 2004; Regli et al., 2004; Huggenberger and Aigner, 1999; Weissmann et al., 1999; Rauber et al., 1998; Carle et al., 1998; Fogg et al., 1998; Deutsch and Wang, 1996; Webb and Anderson, 1996; Koltermann and Gorelick, 1996; McKenna and Poeter, 1995; Jussel et al., 1994; Webb, 1994; Bridge, 1993; Ashmore, 1993; Paola et al., 1992; Heller and Paola, 1992; Brierley, 1991; Anderson, 1989; Miall, 1985; Ashmore and Parker, 1983; Allen, 1978, Miall, 1978). However, for reasons of complexity of the aquifer structures in these coarse-grained sediments and effects on hydraulic property distribution, stochastic modeling was rarely applied to practical problems (Dagan, 2002). The literature only provides a few examples of investigations conducted on adequately instrumented field sites and related risk analysis (e.g. river-groundwater interaction, Rhine/Wiese sand and gravel aquifer, Switzerland, Regli et al., 2003; bacterial and virus transport and attenuation processes in Dornach, Munich gravel plain aquifer, Germany, Flynn, 2003; hierarchical geostatistics and multifacies systems, Boise Hydrogeophysical Research Site, Idaho, USA, Barrash and Clemo, 2002; physical scale modeling of the Ashburton River gravels, Canterbury Plains, New Zealand, Ashworth et al., 1999, 2004; flow and contaminant transport in quaternary gravel deposits, Steisslingen, Germany, Klingbeil et al., 1999; sand and gravel pit in Stoughton, Wisconsin, USA, Anderson et al., 1999; field study of dispersion in a heterogeneous aquifer, Columbus Site, Mississippi, USA, Boggs et al., 1992; methodologies for groundwater driven health risk assessment in heterogeneous aquifers, Maxwell and Kastenberg, 1999, Maxwell et al., 1998; environmental research field site Horkheimer Insel, Germany, Teutsch and Kobus, 1990; various studies on flow and transport in a sand aquifer, Borden site, Canada, Sudicky, 1986, Mackay et al., 1986).

The generally scarce information on outcrop and the existing buildings and infrastructures preventing high resolution Ground Penetrating Radar (GPR) are important reasons for not considering heterogeneity in practical applications, particularly in urban areas. Therefore, drill-core descriptions with the known disadvantages concerning data quality and sedimentary structure recognition (Regli et al., 2002) are the only sedimentological data available. In urban areas, large amounts of geological and hydrological data are generally available but spread in different institutions. Since localizing this data is often difficult and its preparation for specific questions time-consuming, a Geological Database (GeoData) for northwestern Switzerland was set up (Kirchhofer, 2006). It comprises a systematic data collection, analysis of drill-core data, and assessment of metadata from geological and hydrological reports. GeoData can be linked to a Geographic Information System (GIS) to provide together with groundwater head and further hydrological data, a unique data source suitable for empirical studies and hypothesis testing in the field of quantitative information on urban hydrological questions. A method of combined sedimentary structure and geostatistical analyses of drill-core data was presented by Regli et al. (2002). The results of the analyses are used to develop various simulations of stochastically generated aquifer properties, which can subsequently be integrated into multilayer high-resolution groundwater models. The investigations center on determining to which extend available drill-core descriptions may be used to describe heterogeneity in coarse river systems and how the different strategies influence the understanding as well as prediction of the impacts and changes on groundwater flow regimes.

This study describes subsurface heterogeneity investigations during groundwater management of a subsurface highway construction (Fig. 1). During single construction phases, considerable groundwater drawdown was necessary, leading to significant changes in the groundwater flow regime. The effect of open sump drainage during construction of an emergency exit is presented. Furthermore, a comparison was established of the changes observed before and after this construction phase. The three investigated situations reveal considerable differences in hydrological and operational boundary conditions and allow evaluating the influence of alternating boundary conditions against the various aquifer heterogeneity simulations. The results are also compared in the context of the varying boundary conditions assuming: (1) uniform distribution of equivalent aquifer parameters and (2) heterogeneous distribution of aquifer parameters resulting from sequential indicator simulations based on drill-core data. The results obtained are discussed on the basis of groundwater budgets, flow paths and flow velocities.

Groundwater protection as well as policy and management aspects should already be considered at the early stages of urban planning to reconcile the various individual and often conflicting interests. The main objectives of engineers and constructors are to complete construction with minimal effort as regards guaranteed drawdown levels, inexpensive groundwater disposal and safeguarding nearby groundwater use. The environmental and civil engineering offices involved rather focus on possible mobilization of contaminants and changes in the groundwater flow regime during and after construction. However, to achieve a sustainable development of urban groundwater resources, the sum of all impacts on groundwater flow regimes should be taken into account. The investigations focus on the applied aspects of data fusion within the context of subsurface highway construction.

2 Settings

2.1 Hydrogeological Setting

Basel, located in northwestern Switzerland, borders both Germany and France. The Rhine enters Basel from the east and changes its course at an angle in northern direction (Fig. 1). The highway construction sections outlined in this book chapter are located in the northern part of Basel to the left of the river Rhine.

The shallow unconfined aquifer mainly consists of late Pleistocene gravel deposited by the Rhine. The gravel deposits, interbedded with fine-grained, flood plain sediments result in variable conductivity within the aquifer. The thickness of the aquifer ranging between 15 and 35 m is underlain by an

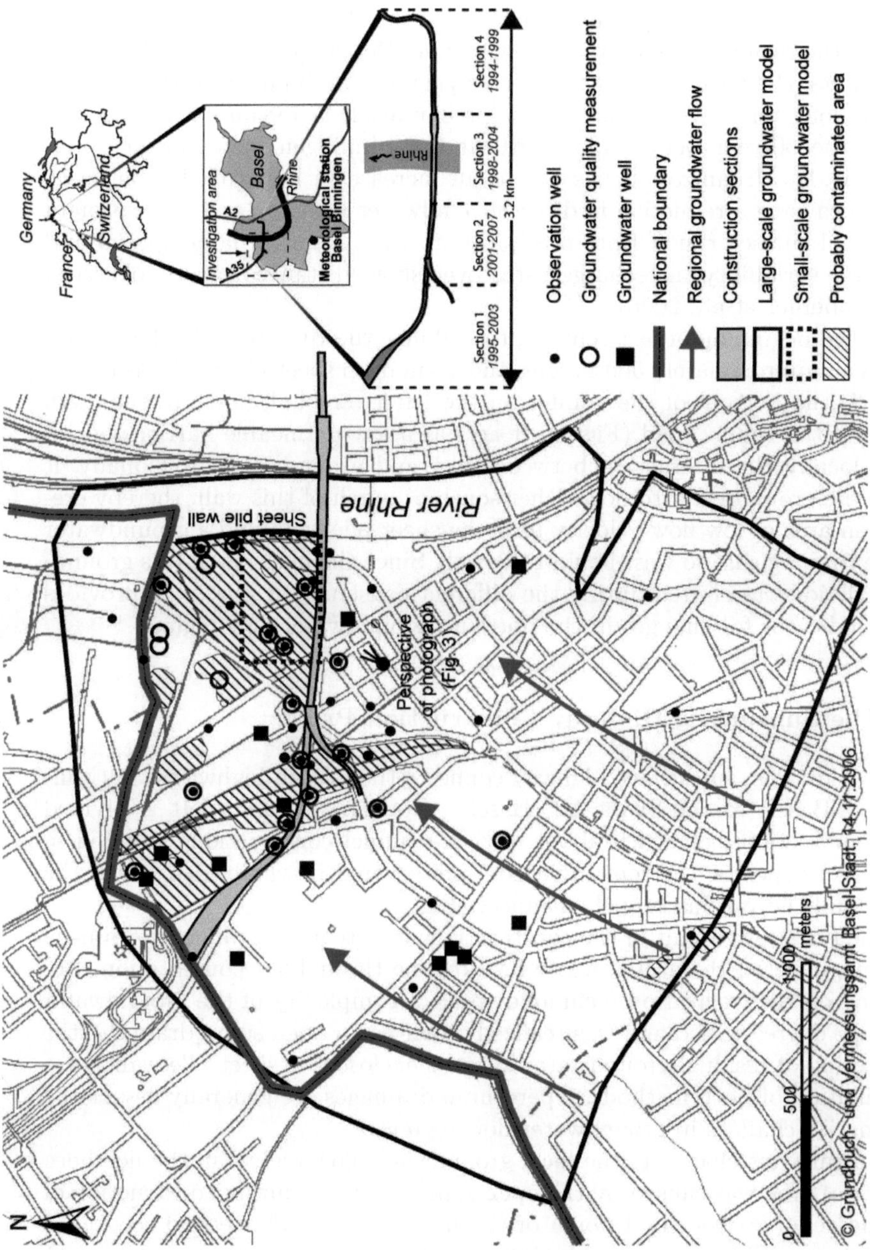

Fig. 1. Investigation area in the city of Basel, northwestern Switzerland. The regional groundwater flows mainly from southwest to northeast. Note that construction of the tunnel highway runs at an approximately 60° angle (CCW) to the regional groundwater flow. Construction years of each section are given in the center of the right column

aquiclude composed of Oligocene mud to clay rich sediments. The general slope and the main direction of the regional groundwater flow are from southwest to northeast.

The river-groundwater interactions along the Rhine are an important element of the regional groundwater flow regime. Depending on the hydrological constraints, the river acts both as a receiving and an infiltrating stream.

The long-term average for yearly precipitation is $788\,\mathrm{mma}^{-1}$, measured during the 30-year period 1961–1990 at the Binningen meteorological station (Fig. 1). Urbanization has led to an increase in impermeable surfaces, thereby causing a reduction in direct groundwater recharge and generation of additional surface runoff from precipitation. As a result, a large spatial and temporal variability in recharge rates over short distances can be observed (Huggenberger et al., 2006).

At the beginning of last century, to stabilize the river bank for harbor facilities an approximately 500-m long and 20-m deep sheet pile wall was driven down to the bottom of the aquifer on the left river bank north of the main course of the tunnel road (Fig. 1). It acts as a low-permeable barrier and reduces locally the interaction between river and groundwater. Regionally, it forces the groundwater to flow either south or north of this wall, thereby creating an area of low flow velocity near the sheet pile wall, and a groundwater divide running east to west behind the wall. Since the position of this groundwater divide was shifted during the different construction phases, it provides a key indicator for changes in the northern groundwater flow regime.

2.2 The Subsurface Highway Construction Project

The 3.2-km long subsurface highway connects the French highway A35 (Mulhouse – Basel) to the Swiss A2 (Basel – Gotthard – Milan). It is divided into four sections, of which about 87% are tunnel constructions situated in the gravel deposits; the remaining 13% are covered by the bridge across the Rhine and the various tunnel entrances (Fig. 1).

The progressive shift of the construction sites, requiring different drainage systems, affected the groundwater flow regime throughout construction. Depending on the excavation technique applied, complexity of the groundwater drainage degree varied and was realized either as an open sump drainage, the dewatering of residual groundwater in areas enclosed by sheet pile walls, or a combination of both methods. Open sump drainages are generally associated with major changes in groundwater flow regimes.

A significant change in the local groundwater flow regime in the northern industrial area was caused by the open sump drainage during construction of an emergency exit located north of the main track in the second section of the highway (Figs. 1 and 2). Maximum extraction rates ranging between 15 and $20\,\mathrm{ls}^{-1}$ are relatively low. Nevertheless, on account of the small hydraulic gradient in the adjacent industrial area to the north, the effect of this drainage

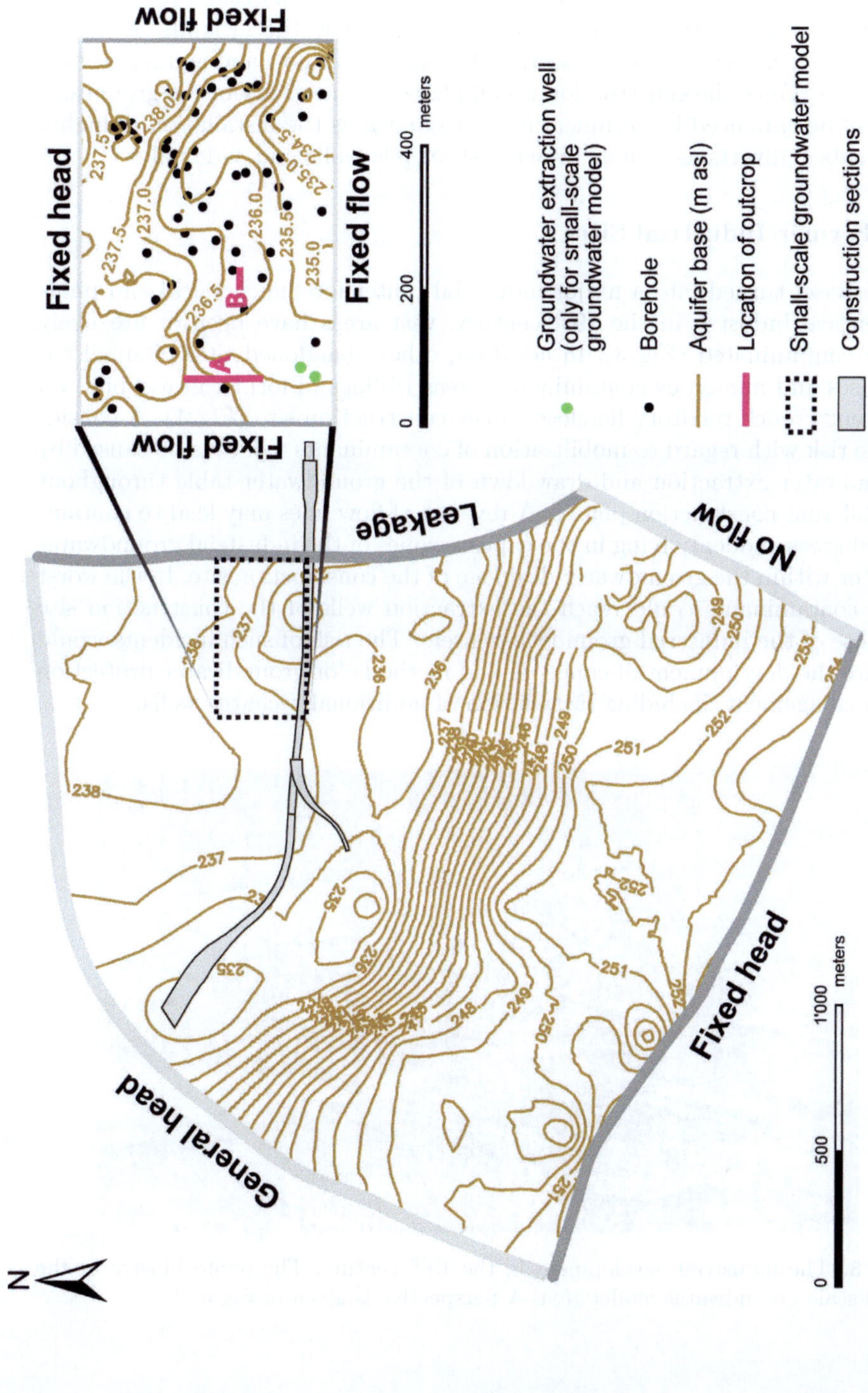

Fig. 2. Surface of the aquifer base and boundary conditions used for large and small-scale groundwater models. Note the steep slope in the center of the model area of the large-scale groundwater model

on the regional groundwater flow regime was relatively large and resulted in a change in groundwater flow paths and a shift of the local groundwater divide.

In some areas, the tunnel is permanently below the groundwater table leading to a reduction of cross-sectional groundwater flow and aquifer-storage capacities. Once the construction is completed, connectivity of the groundwater will be enhanced by technical measures, such as the installation of highly permeable culverts as well as drawing sheet pile walls and slide pales.

2.3 Former Industrial Sites

Since Basel turned into a major industrial center for the chemical and pharmaceutical industry in the 19th century, vast areas have been or are likely to be contaminated (Fig. 3). In addition, other abandoned sites of small enterprises and numerous contaminated areas (fillings of former gravel pits) on adjacent French territory lie close to the construction site (Fig. 1). A considerable risk with regard to mobilization of contaminants will thus be caused by groundwater extraction and drawdown of the groundwater table throughout the different construction phases. A reversal of flow lines may lead to contaminated areas suddenly lying in the capture zones of the industrial groundwater wells or within the groundwater drainage of the construction site. In the worst case, contaminants could reach the extraction wells of the construction site or those of the industrial groundwater users. The risk of such incidents would require the development of concepts and methods for groundwater protection and management, including installation of additional recharge wells.

Fig. 3. The industrial development in the 19th century. The photo illustrates the small-scale groundwater model area. A perspective is given in Figure 1

3 Conceptual Approach and Methodology

The conceptual approach and methodology used to answer a series of questions generally arising in the context of urban subsurface road constructions include the following points: (1) Is the available data sufficient to adequately investigate groundwater flow and transport, and what questions can be answered with the existing data? (2) What additional data could improve predictions? (3) How could data acquisition be optimized?

A regional-scale groundwater model was used, adapted and calibrated regularly during the various construction site drainages and throughout construction. Due to the numerous calibrations (at least two times per year) the prediction capabilities of the groundwater model were continuously enhanced. This facilitated optimizing extraction and recharge rates during construction site drainages and detecting changes in the groundwater flow regime. Aquifer parameters obtained from pumping tests are therefore of prime importance. However, to adequately model groundwater transport, further knowledge of subsurface heterogeneity and the distribution of aquifer parameters are essential. This knowledge may to a certain degree be derived from drill-core information, sedimentary structure analysis at outcrops and from geophysical as well as hydraulic measurements.

The conceptual approach is given in Figure 4. The investigated area was delineated and comprises an inventory of all relevant boundaries characterizing the regional and local groundwater flow regime, including all possible impacts. The following three-step approach for data fusion was chosen: (1) determination of data requirements; (2) data processing and finally (3) data evaluation. The presented example of data fusion first identified and collected the required data and subsequently formulated the basic requirements for setting up groundwater models with accurate boundary conditions. The data required comprises hard and soft data. Hard, reliable data is derived for example from outcrops and groundwater head measurements. Soft data is obtained from drill-core descriptions and groundwater quality analyses. The first element of the groundwater management system at this level of data fusion consists in: (i) a dense groundwater-monitoring network where hydraulic heads are measured and groundwater chemistry is analyzed regularly. The data obtained is then processed in a second step. The second main element of the groundwater management system comprises: (ii) a database together with the development of an application facilitating lithofacies-based interpretation of drill-core data. Export of this information can directly be deployed in the third element of the groundwater management system consisting of: (iii) geostatistical analyses of the aquifer heterogeneity leading to conditioned stochastic aquifer simulations. The fourth element of the groundwater management system consists in: (iv) regional and local high-resolution groundwater modeling. The geostatistical simulations are integrated into the local groundwater models. The third and last step of data fusion finally comprises an evaluation of the processed data. Assisted by a GIS, the simulation results are visualized to

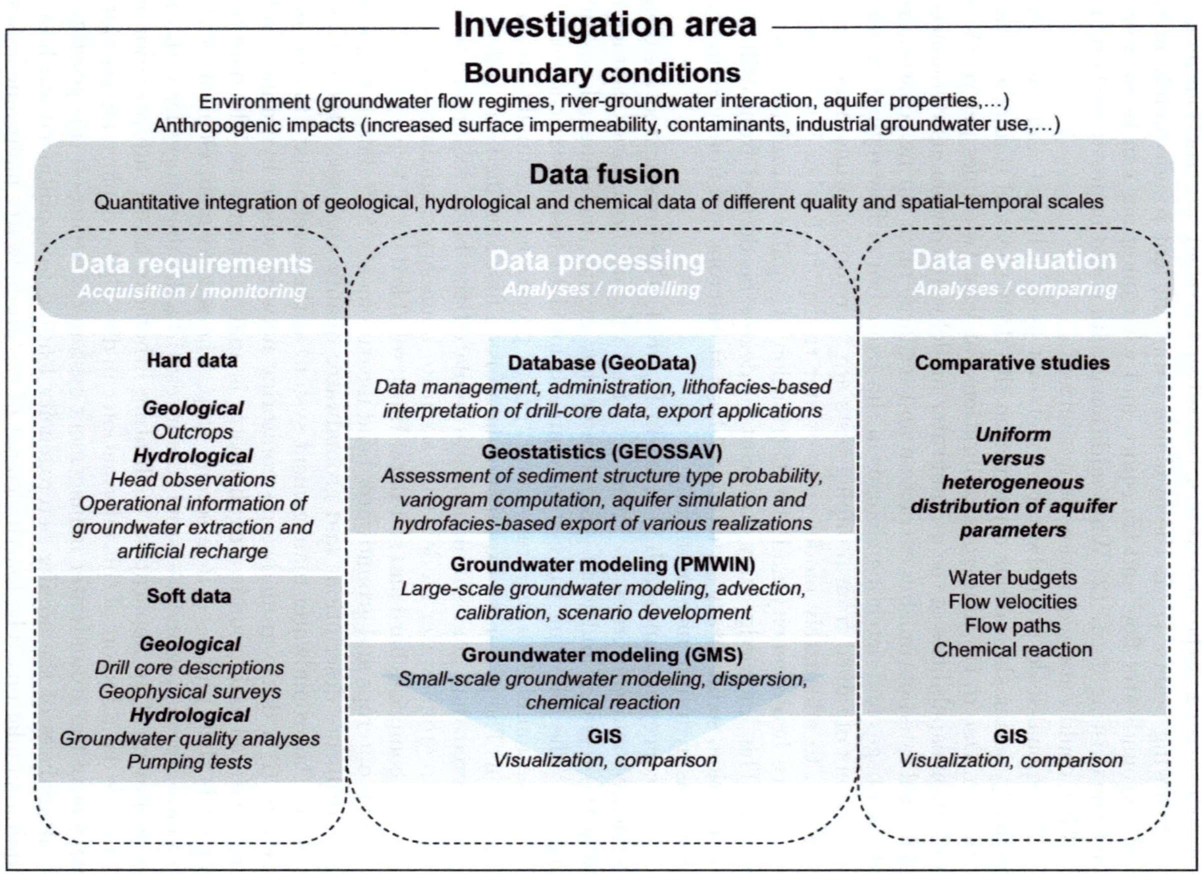

Fig. 4. Conceptual approach of data fusion for practical urban hydrogeological applications

improve the understanding among engineers and all other stakeholders. The four elements are described in detail in the following text.

3.1 Groundwater Monitoring Network

The network comprises a total of 44 observation wells instrumented by automated water-level loggers for continuous measurement of the hydraulic head (Fig. 1). The hydrographs of this observation network are analyzed monthly. A total of 21 observation wells are sampled regularly for groundwater quality measurements. Furthermore, the extracted water for industrial groundwater use and for settling tanks on the construction sites is sampled at regular intervals. The monitoring program was adapted to the progress of the various construction sections, to the current groundwater management requirements and to the results obtained from groundwater modeling. Interpretation of the changes observed in groundwater quality measurements together with the modeling results allowed optimizing the localization of new observation wells.

3.2 Geological Database

GeoData (Kirchhofer, 2006) of the city of Basel includes some 3,000 drill-core descriptions. It can be used for almost any question arising during setup and operation of other elements of the groundwater management system. Application of GeoData for calculation of the aquifer base, evaluation of aquifer parameters and for lithofacies-based interpretation of drill-core data is summarized hereafter.

Calculation of Aquifer Base and Evaluation of Aquifer Properties

The aquifer base in the large-scale groundwater model includes information on more than 400 drill-cores, whereas the small-scale groundwater model includes information on 71 drill-cores (Fig. 2).

Distribution of horizontal conductivity zones in the large-scale groundwater model was based on different type and quality data sets available from GeoData including various pumping tests. Furthermore, additional information from reports outlining regional geological questions, including pumping test data, was used to determine hydraulic parameter distribution.

Combining Outcrop with Drill-Core Data for Sedimentary Structure Analyses

Information on the architecture of the aquifer is required to adequately model subsurface heterogeneity. Outcrop and drill-core information contains data of different quality and resolution at different scales. Outcrops of natural deposits above the groundwater table reveal distinct and coherent structural elements such as lenses and layers of different gravel types (Jussel et al., 1994; Siegenthaler and Huggenberger, 1993; Huggenberger et al., 1988). Outcrop investigations therefore provide extremely reliable hard data. However, outcrops are restricted to one large excavation pit north of the investigation area and, to some extent, to outcrops within the tunnel itself. Due to easy access and visibility, undisturbed sedimentary structures and textures at the outcrops could be examined in detail and the sedimentary structure patterns of the investigation site defined. Definition of sedimentary texture types is based on grain-size distribution and sediment sorting. Sediment structure types are made up of one or a combination of two possibly alternating sediment texture types.

Drill-core data provides limited information on the spatial distribution of sedimentary structures and subsurface properties such as hydraulic conductivity, porosity and dispersivity. Drilling may destroy sedimentary structures and smear the interface with adjacent layers. Typically, drill-core layer descriptions are not very detailed and do not clearly indicate explicit sedimentary structure types. Furthermore, the quality of individual drill-core descriptions varies considerably depending on the geotechnical or sedimentological approach used, thus permitting limited and speculative conclusions on sedimentary structures. Results on a drill-core scale have to be evaluated carefully, as fractures common in a drill-core may not clearly reveal the overall flow on a field scale for lack of interconnectivity or dominance of high permeable porous sedimentary structure types. Breaks in cores may also be attributed to core drilling and handling processes. Consequently, drill-core data is regarded as soft data. The concept of determining sedimentary texture and structure types is based on the concept developed for the Rhine gravel as described by Rauber et al. (1998), Jussel et al. (1994), Siegenthaler and Huggenberger (1993). The interpretation method of drill-core data is elucidated in Regli et al. (2002).

Sedimentary structure types of this area include: open-framework gravel (OW), open-framework/bimodal gravel couplets (OW/BM), gray gravel (GG), brown gravel (BG), alternating gray and brown gravel layers (GG/BG), horizontally-layered or inclined, silty gravel (SG), sand lenses (SA), and silt lenses (SI). Regli et al. (2002) and Huggenberger and Regli (2006) give a detailed description of the sedimentary structure types.

Sedimentological drill-core descriptions of coarse gravel deposits provide information on composition and texture of deposits. More specifically, details of grain-size categories, sorting, composition of major constituents and proportion of each grain-size fraction can be determined and information obtained on color, chemical precipitation, thickness of a deposit, and its transition through the underlying layer. The quality of the descriptions varies considerably. Important sedimentary structure types, such as the highly permeable OW, are generally overlooked due to smearing with overlying and underlying layers during the drilling process. Occurrence and size of OW determine, however, variance and correlation length of the hydraulic conductivity in coarse gravel deposits (e.g. Jussel et al., 1994). Consequently, an important gap exists between outcrop and drill-core descriptions. The strong association of OW to the related structure type OW/BM has led to the concept of a gradual sedimentary structure-based interpretation of outcrop, drill-core and GPR data. The method presented by Regli et al. (2002) allows a probability assessment of drill-core layer descriptions representing defined sedimentary structure types.

Based on drill-core information, defined sedimentary structure pattern for the investigation site and on the interpretation method for drill-core data, GeoData allows analysis of sedimentary structure and provides data sets of point information with arbitrary separation distances along drill-cores. The point information includes space data (x-, y-, z-coordinates), probabilities of sedimentary structure types (probability that a drill-core layer description represents a defined sedimentary structure type) and an indication of the most likely sedimentary structure type (Fig. 5).

3.3 Geostatistics

Modeling spatial data variability is the key to any subsurface simulation. A variogram describes the spatial correlation of data as a function of the separation vector h between two data points. Computation of the indicator variogram was based on the drill-core data. Indicator transformation at grid node locations was set at unity for sedimentary structure types $s_k, k = 1, \ldots, K$, with the highest probability values; otherwise it was set at zero (Deutsch and Journel, 1998):

$$i(u_\alpha; s_k) = \begin{matrix} 1, \text{if } s(u_\alpha) = s_k \\ 0, \text{otherwise,} \end{matrix} \tag{3.3.1}$$

where u_α refers to a particular data location and $s(u_\alpha)$ to a particular data value.

Experimental indicator variograms were calculated after Eq. 3.3.2 for various directions (azimuth, dip, plunge) using GEOSSAV (Regli et al., 2004), an interface to selected geostatistical programs available from the Geostatistical Software LIBrary, GSLIB (Deutsch and Journel, 1998):

$$\gamma_I(h; s_k) = \frac{1}{2N(h)} \sum_{1}^{N(h)} \left[i(u; s_k) - i(u+h; s_k)\right]^2, \tag{3.3.2}$$

where $N(h)$ is the number of data pairs, $i(u; s_k)$ is the indicator at the start or tail of the pair, and $i(u+h; s_k)$ the corresponding end or head indicator.

To derive maximum benefit from available textural and structural information, the aquifer structure was derived by sequential indicator simulation. For a specific category, the indicator kriging estimate, i.e. the probability that s_k prevails at location u, is written as a linear combination of the n nearby indicator-coded data (Deutsch and Journel, 1998):

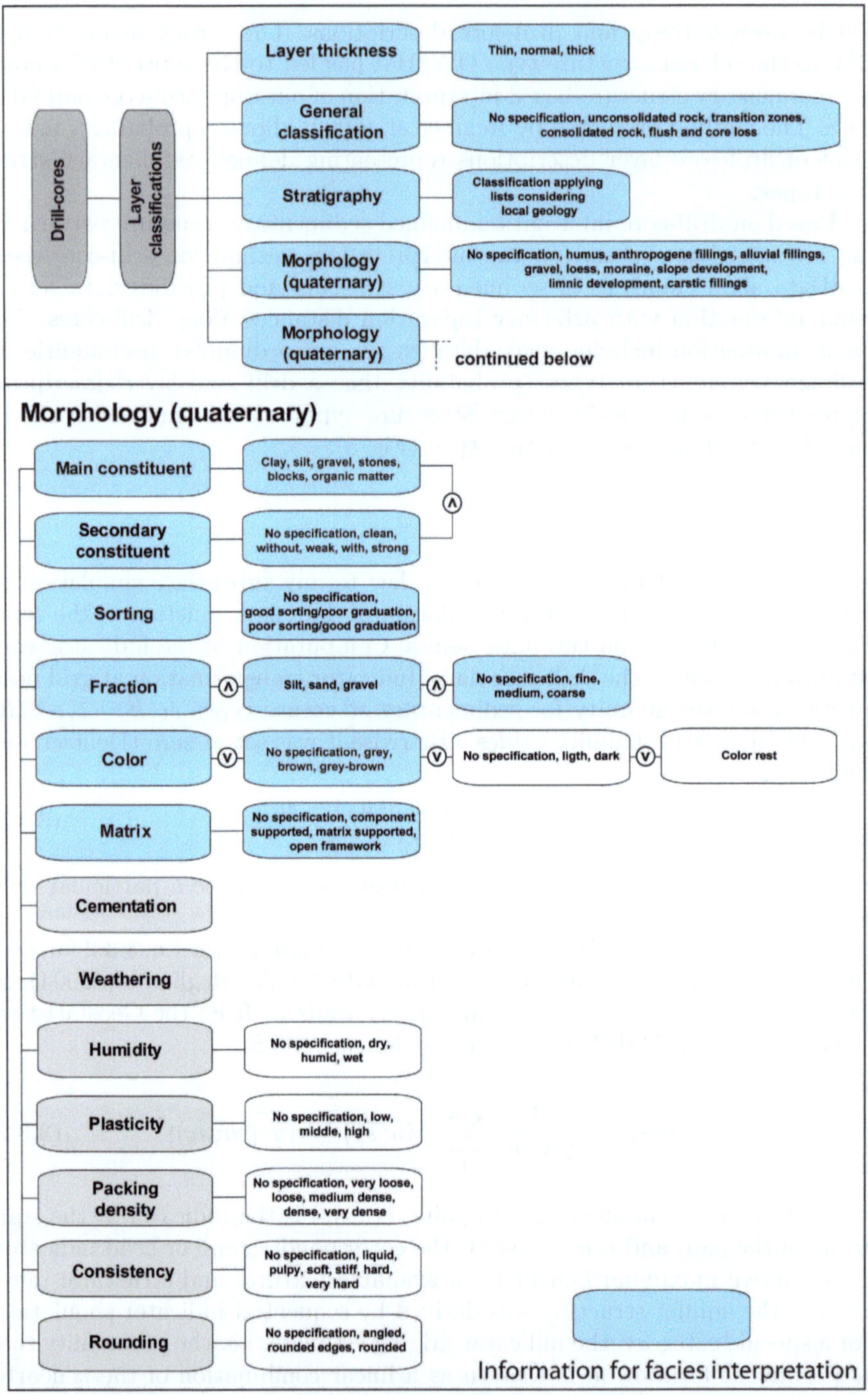

Fig. 5. Drill-core input mask for GeoData (∨: or; ∧: and). The information used for sedimentary structure interpretation is highlighted

$$
\begin{aligned}
[i(u;s_k)]^*_{SK} &= [Prob\{S(u)=s_k|(n)\}]^*_{SK} \\
&= \sum_{\alpha=1}^{n}\lambda_\alpha(u;s_k)\,i(u_\alpha;s_k)+\left[1-\sum_{\alpha=1}^{n}\lambda_\alpha(u;s_k)\right]F(s_k)
\end{aligned}, \quad (3.3.3)
$$

where $F(s_k)$ is the stationary prior probability of category s_k, and the $\lambda_\alpha(u;s_k)$'s are the indicator kriging weights corresponding to category s_k, which depend on the closeness of the data considered for the estimation.

The sequential indicator simulation principle is an extension of conditioning and comprises all data available within the neighborhood of a model cell, including the original data and all previously simulated values. Sequential indicator simulations are processed by a number of steps. An initial step establishes a grid network and coordinate system. This is followed by assigning data to the nearest grid node. Where more than one data point may be used at a node, the closest data point is assigned to the grid node. In a third step, a random path through all grid nodes is determined. For a node on a random path, adjacent data and previously simulated grid nodes are searched to allow assessment of the conditional distribution by indicator kriging. Based on this distribution, a simulated sedimentary structure is randomly drawn and set as hard data before selecting the next node in the random path prior to repeating the process. By using this approach, the simulation grid is sequentially built up. During the final step of the sequential indicator simulation, results are checked to ensure that orientations and sizes of the simulated sedimentary structures concur with those observed.

For groundwater modeling, the simulated sedimentary structure needs to be transformed into hydraulic parameters. The generated sedimentary structures are characterized by average and randomly selected hydraulic conductivity and porosity values provided by average and standard deviations calculated by Jussel et al. (1994) (Table 1). Files containing distributions of

Table 1. Hydraulic parameters of sedimentary structure types used in the characterization of aquifer simulations (after Jussel et al. 1994). Refer to chapter 3.2 for the abbreviations.

	Sedimentary structure type								
	OW	OW/BMGG		BG	GG/BG horizontal	GG/BG inclined	SG	SA	SI
Hydraulic conductivity K $[\mathrm{mms^{-1}}]$	100	10	0.15	0.02	0.08	0.1	0.008	0.26	0.005
Standard deviation $\sigma_{\ln K,1}[-]$	0.8	0.8	0.5	0.6	0.8	0.8	0.5	0.4	0.4
Porosity n [%]	34.9	30	20.1	14.1	17	17	25	42.6	40
Local longitudinal dispersivity [mm]	25	30	25	30	30	30	3	0.3	0.05

hydraulic conductivity, effective porosity and dispersivity values were generated and exported to groundwater models. According to sedimentological and geostatistical analyses of the aquifer, each aquifer simulation corresponds to various equiprobable representations of the subsurface at variable degrees of uncertainty in hydraulic parameter values and geometry of the sedimentary structures.

3.4 Groundwater Modeling

Regarding selection and setup of an appropriate groundwater model, it is important to ensure that the chosen model and desired resolution are capable of answering relevant questions simultaneously. When applying groundwater models, one of the key requirements is high-quality, site-specific data (National Research Council, 1990). Before setting up a groundwater model, groundwater budgets and boundary conditions – the main components of a groundwater system – must be identified and analyzed.

A large and a small-scale groundwater model combination were chosen for this case study (Fig. 1 and Fig. 2). While the large-scale groundwater model was used to simulate the regional groundwater flow regime throughout the entire construction period of the tunnel road, analyses of aquifer heterogeneity were integrated into the small-scale groundwater model. For selected construction phases, the large-scale groundwater model allowed to define boundary conditions for the telescoped small-scale groundwater model. Both large and small-scale groundwater models were simulated using the 3D finite difference code MODFLOW (McDonald et al., 2000):

$$div\left(K \cdot \vec{\nabla} h\right) + W = S_s \frac{\partial h}{\partial t}, \tag{3.4.1}$$

where K, are values of hydraulic conductivity along the x-, y-, and z-coordinate axes, which are assumed to be parallel to the major axes of hydraulic conductivity (L/T); h is the hydraulic head (L); W is a volumetric flux per unit volume representing sources and/or sinks of water, with $W<0.0$ for flow out of the groundwater system, and $W>0.0$ for flow in (T-1); S_s is the specific storage of the porous material (L-1); and t is time (T). Eq. 3.4.1, when combined with boundary and initial conditions, describes three-dimensional groundwater flow in a heterogeneous and anisotropic medium, provided that the principal axes of hydraulic conductivity are aligned with the coordinate directions. The groundwater flow process solves Eq. 3.4.1 using the finite-difference method in which the groundwater flow system is divided into a grid of cells. Further, the solver package PCG2 (preconditioned Conjugate Gradient) with modified incomplete Cholesky preconditioning was applied (Hill 1990).

The graphical user interface Processing Modflow (PMWIN) of Chiang (2005) was used for the large-scale groundwater model. The graphical interface

Groundwater Modeling System (GMS) of Environmental Modeling Systems Inc. (2006) was applied to the small-scale groundwater model. Two different graphical user interfaces were applied, as the large-scale model has already been in operation with PMWIN since 2001, and the telescoped, small-scale model has recently been operated with GMS.

Large-Scale Groundwater Modeling

The large-scale groundwater model covers an area of $2,720\,\text{m} \times 2,860\,\text{m}$ (about $8\,\text{km}^2$; Fig. 1 and Fig. 2). The spatial discretisation resulted in cell sizes varying between $5\,\text{m} \times 5\,\text{m}$ (near the construction site) and $30\,\text{m} \times 30\,\text{m}$ in totally 132,500 cells. An approach with four horizontal layers was chosen to vertically integrate the construction. Construction itself was integrated either as inactive cells or as horizontal flow barriers with defined hydraulic conductivities. During construction, progressive adjustments were made. The surface of the aquifer base and distribution of horizontal hydraulic conductivity zones were derived from different type and quality data sets available from GeoData.

The hydraulic conductivity values range between $3.1\text{E}{-3}\,\text{ms}^{-1}$ and $1.3\text{E}{-4}\,\text{ms}^{-1}$. A 10:1 ratio between horizontal and vertical hydraulic conductivity was chosen. Since the southern part of the model area has a broad steep slope in the aquifer base without any detailed geological information, hydraulic conductivity had to be estimated (Fig. 2). For all model runs, divergence of calculated and observed hydraulic heads is highest in this part and is in the order of 1 m. However, the divergence for the remaining hydraulic heads averages in 0.2 m.

Based on a one-day test measurement of groundwater levels (Wagner et al., 2001) model boundary conditions are of the first type (fixed head) along the southern side, and of the third type (leakage) along the Rhine. The western and northern boundaries are specified as no flow in a first phase and as general head in a second phase, thus resulting in a groundwater outflow south and a groundwater inflow north of the steep slope in the aquifer base. Hydraulic conductivity of the riverbed was set at $5.0\text{E}{-5}\,\text{m}^2\text{s}^{-1}$.

A total of 48 extraction wells were integrated, i.e. nine production wells and one recharge well for industrial groundwater use, 35 wells extracting groundwater along the various construction sections as well as three recharge wells. The hydraulic head was continuously monitored in totally 44 observations wells. As a routine procedure, the groundwater model was calibrated at least biannually.

Small-Scale Groundwater Modeling

The small-scale groundwater model covers an area of $450\,\text{m} \times 300\,\text{m}$ ($0.135\,\text{km}^2$; Fig. 1 and Fig. 2). The spatial discretisation resulted in totally 1,350 cells of $10\,\text{m} \times 10\,\text{m}$ cell size. For an appropriate vertical integration of aquifer heterogeneity, an approach with 13 horizontal layers was chosen. Each layer is 2-m thick and the total maximum vertical thickness amounts to 26 m. The interpolated surface of the aquifer base was then cut with the model grid, thus resulting in partly inactive cells in the two lower model layers. Finally, the interpolated aquifer properties (hydraulic conductivity, effective porosity and longitudinal dispersivity) were assigned to the prepared grid.

The large-scale groundwater model provided the boundary conditions. Model boundary conditions are of the first type (fixed head) along the northern side, and of the second type (fixed flow) along the eastern, southern and western side to account for the variable inflow and outflow rates across the boundaries. The eastern side covers residual leakage of the Rhine through the sheet pile wall, the southern side the groundwater flow beyond the tunnel construction and the western side focuses on the regional groundwater flow through this area.

Since hydraulic conductivity varies with scale of measurement, it can increase to over several orders of magnitude. As larger blocks of the subsurface are tested for subsurface flow, preferred pathways are encountered that increase the measured average hydraulic conductivity value (Carrera, 1993). The preferred pathways are provided by sedimentary structure heterogeneities, fractures or flow conduits (Schulze-Makuch et al., 1999).

Distribution of hydraulic conductivity for the large-scale groundwater model is derived from a series of pumping tests describing the average hydraulic conductivity for this region. A 2.0E$-$3 ms^{-1} average value obtained from this distribution is also used for the small-scale groundwater model of uniform aquifer parameter distribution. Together with the median from the geostatistical analyses, the scaling factors obtained for hydraulic conductivity range between 13 and 20. Furthermore, the scaling behavior by an empirical power law of the hydraulic conductivity proposed by Schulze-Makuch et al. (1999) was employed:

$$K = 10^{c} \, (V)^{m} \,, \tag{3.4.2}$$

where K is the hydraulic conductivity, c is a parameter characteristic for a geological medium that relates to geological variables such as average pore size and pore interconnectivity in porous media, V is the volume of tested material (used as scale measure), and m is the scaling exponent (slope of the line on the log-log plot). For unconsolidated media and alluvium parameters, for c and m values of -4.8 and 0.5, respectively, are proposed. The volume taken account of in this case study for the small scale groundwater model amounts to approximately 3.5E6 m^3. This results in a calculated hydraulic conductivity value of 3.7E$-$3 ms^{-1}, which is in the same order of magnitude as the average hydraulic conductivity of 2.0E$-$3 ms^{-1} used in the large scale groundwater model. When applied to the present study, the scaling factors range between 25 and 37. Finally, a decimal power-scaling factor 10 was chosen and a 10:1 ratio between horizontal and vertical hydraulic conductivity.

Table 2. Water budgets across model boundaries, extraction rates of construction site drainage and altitude of the specified head in the north. Flows are considered "in" if they are positive and "out" if they are negative.

	Flow budgets [m^3d^{-1}]				Specified head [m asl]
	East	South	West	Wells	North
March 23, 2003	−198.7	−691.2	+423.4	0.0	245.54 – 244.88
October 28, 2004	+60.5	+380.2	+691.2	−604.8	245.20 – 244.96
February 16, 2006	+34.6	+120.96	+155.2	0.0	244.71 – 244.56

The value for effective porosity is the determining parameter for the advective processes and the time scale of contaminant transport in the aquifer (Graf and Schäfer, 2002). Porosity may also vary with scale, however, such variations are assumed to be minimal compared to those encountered for hydraulic conductivity and generally ranging over several orders of magnitude (Schulze-Makuch et al., 1999). Evaluations reveal that the 0.2 value assumed for the uniform distribution of porosity from the large-scale groundwater model is of the same order of magnitude as the value for the various aquifer heterogeneity simulations. In unconfined aquifers, the value for the specific yield is the same as the one for effective porosity. An empirical value of 1.0E−4 m^{-1} for sandy gravel was chosen for specific storage (Anderson and Woessner, 1992).

Simulations were conducted for three sets of hydrological and operational boundary conditions (Table 2): Set 1 considers boundary conditions for an ordinary situation, preceding the major drawdown phases (March 23, 2003). Set 2 covers the period with the most significant changes in the local groundwater flow regime north of the industrial area and caused by the open sump drainage during construction of an emergency exit located north of the main track in Section 2 (October 28, 2004). Set 3 describes again boundary conditions for an ordinary situation and accounts for low groundwater levels and Rhine infiltration after the major drawdown phase (February 16, 2006).

Flow Velocities

GMS allows determining groundwater flow velocity and seepage velocity. The velocity calculator uses three scalar data sets from which it creates a vector data set. Head, porosity and hydraulic conductivity are the three input data sets used. Furthermore, a vertical anisotropic factor has to be chosen. Darcy's law is applied to the calculations:

$$v_s = \frac{v_d}{n} = \frac{ki}{n}, \tag{3.4.3}$$

where v_s is the seepage velocity, v_d is the Darcy velocity, n is the effective porosity, k is the hydraulic conductivity, and i is the head gradient.

In 3D, the equation is:

$$div\left(v + K \cdot \vec{\nabla} h\right) = 0, \tag{3.4.4}$$

where K, are values of hydraulic conductivity along the x, y, and z coordinate axes, which are assumed to be parallel to the major axes of hydraulic conductivity (L/T); h is the hydraulic head (L). To calculate the hydraulic gradient vector the finite differences are computed by:

$$\frac{\partial h}{\partial x} = \frac{\frac{h_{i_{+1}jk} - h_{ijk}}{x_{i_{+1}jk} - x_{ijk}} + \frac{h_{ijk} - h_{i_{-1}jk}}{x_{ijk} - x_{i_{-1}jk}}}{2}. \tag{3.4.5}$$

The units of the calculated flow velocities are length over time (L/T).

4 Results

4.1 Outcrop and Drill-Core Analyses

The outcrop within the perimeter of the former industrial sites above the groundwater table (Fig. 6) clearly reveals a prevalence of poorly sorted, clean gravel without significant fine sediment fraction on top of the sequence. Below, sets of gravel couplets were observed comprising OW and OW/BM interspersed with thin layers of poorly sorted sometimes sandy clean gravel. The Pleistocene Rhine gravel along the upper Rhine valley generally exhibits a higher amount of high permeable structure types in the lowest part of the sequence. Based on the preservation potential of different sedimentary structure types (Huggenberger and Regli, 2006) it is assumed that the gravel below the groundwater table most likely follow this trend.

Separation to generate point information for sedimentary structure analyses was set at 0.5-m distance. Therefore, a total of 1,999 data points were generated for all drill-cores. The fraction of sedimentary structure types, which is derived in percentage, represents the initial probability density functions for stochastic aquifer simulation (Table 3).

In relation to overbank deposits composed of SA and SI (see Fig. 6A), the large GG and BG fractions are interpreted as products of significant sediment aggradation during lower channel mobility of the former braided river system. In fact, the last landscape-shaping flood events in this area date back 5,000 to 6,000 BC (Rentzel, 1994). Scour and trough fill deposits, mainly consisting of sets of OW/BM couplet cross-beds (see Fig. 6B), are readily found in outcrops. However, the percentage fraction in drill-cores appears to be highly underestimated.

The results of relative sedimentary structure types of the present study are compared with those of Jussel et al. (1994) and Regli et al. (2004) within the context of the fluvial systems, methodological aspects and observed sedimentary structures at the outcrops of the individual investigation sites (Table 4).

Fig. 6. Sedimentary structures of Pleistocene Rhine gravel in Basel. A: flooding dominated sequences in a section approximately parallel to the former flow direction. B: trough-fill dominated sequences approximately perpendicular to the former flow direction (see Fig. 2 for location of outcrops). Refer to chapter 3.2 for the abbreviations

Table 3. Parameters used for the sequential indicator simulation to define the geometric anisotropy of the sedimentary structure types. Refer to chapter 3.2 for the abbreviations. Values in italics are estimates; the isotropic nugget constants of the sedimentary structure types are zero; the variogram models of the sedimentary structure types are exponential; the dip and plunge of the sedimentary structure types are established at zero degrees.

	Sedimentary structure type								
	OW	OW/BMGG	BG		GG/BG horizontal	GG/BG inclined	SG	SA	SI
Relative fraction amount [-]	0.004	0.05	0.49	0.26	0.01	0.03	0.12	0.04	0.002
Azimuth [°]	300	270	320	300	300	280	330	260	300
Maximum horizontal range [m]	2.5	16	9	16	20	20	14	4.3	4
Minimum horizontal range [m]	1.5	5	9	15	10	15	8	3.2	2
Vertical range [m]	0.1	2.5	1.5	1.5	1.5	1.5	2.5	0.6	0.1

Table 4. Comparison of relative fraction amounts of different sedimentary structure types in different case studies. Refer to chapter 3.2 for the abbreviations; [a] Jussel et al. (1994), average distribution of sedimentary structure types in fluvio-glacial deposits of ten gravel pits located in the Rhine, Reuss, Aare, Limmat, and Thur valleys (northeastern Switzerland); [b] Regli et al. (2004), GPR data interpretation in the area of the confluence of the rivers Rhine and Wiese (northwestern Switzerland).

	Sedimentary structure type								
	OW	OW/BM	BG		GG/BG horizontal	GG/BG inclined	SG	SA	SI
Current case study	0.004	0.05	0.49	0.26	0.01	0.03	0.12	0.04	0.002
Jussel et al.[a]	0.028	0.053	0.095	0.158	0.577	0.044	–	0.05	0.004
Regli et al.[b]	0.02	0.06	0.13	0.05	0.50	0.05	0.03	0.04	0.01

The investigation site of the present study is located distal of aggradational areas of the Rhine. The present study focuses mainly on drill-core descriptions and some outcrops analyses. Jussel et al. (1994) investigated average distributions of sedimentary structure types in fluvio-glacial deposits of ten gravel pits located in the Rhine, Reuss, Aare, Limmat, and Thur valleys in northeastern Switzerland. The sites are mainly located proximal to aggradational areas of the rivers, and the investigations centered on outcrop analysis. Regli et al. (2004) examined gravels in the confluence system of Rhine and Wiese in Basel, northwestern Switzerland. The investigations are focused on GPR data interpretation and some drill-cores.

According to Jussel et al. (1994) and Regli et al. (2004), the relative amount of OW is clearly higher than that observed in the present case study. We actually expected a larger amount of OW and OW/BM from the topographic position within the vertical record of the site and from the larger amounts of SA facilitating lateral migration (Church, 2002). Owing to destruction of sedimentary structures and mixing of different units during drilling, we conclude that OW is underestimated for sedimentary structure analysis from drill-core descriptions. The relative amount of OW/BM is similar in all three case studies. This sedimentary structure type indicates trough structures. The larger GG amount may be explained by GG dominance near the top of the vertical section caused by the Rhine abandoning this area in 6,000 BC and subsequent lack of river dynamics. Compared to data from outcrops and GPR, differentiation is considerably more difficult between GG, BG, GG/BG-horizontal and GG/BG-inclined in the various drill-cores of the present study. The sums of the relative amounts of these different sedimentary structure types are, however, of similar range in all case studies. The fact that the GG fraction predominates in the present case study is attributed to predominating aggradational processes or at least to lower fluvial system dynamics. This is supported by the theory that proximal regions and confluence systems are highly energetic and dynamic fluvial systems. Compared to the other two investigations, the relatively higher amount of fine-grained structure types, such as SG, SA and SI in the present case study, indicates slightly large-scale inundations of the entire floodplain.

4.2 Geostatistical Analyses

Variogram Computation

Table 3 includes the resulting parameters of variogram computation for the nine sedimentary structure types identified at the investigation site. Azimuth (both dip and plunge are zero degrees) and the ranges corresponding to maximum and minimum horizontal and vertical spatial correlation distances (Journel and Huijbregts, 1989) characterize the geometric anisotropy of the sedimentary structure types. The results of the variogram analyses, providing orientation of the sedimentary structures, represent the main flow direction of the Rhine during sediment aggradations. The relatively wide spatial correlation ranges from a few meters up to a few tens of meters for the different sedimentary structure types, may be greatly influenced by the density of the data sampled from the drill-core information.

Aquifer Simulation

A sequential indicator simulation is presented in Figure 7. The regular model grid is defined by 45 × 30 × 13 cells of 10 m × 10 m × 2 m cell size. Each simulated sedimentary structure distribution is termed an aquifer realization.

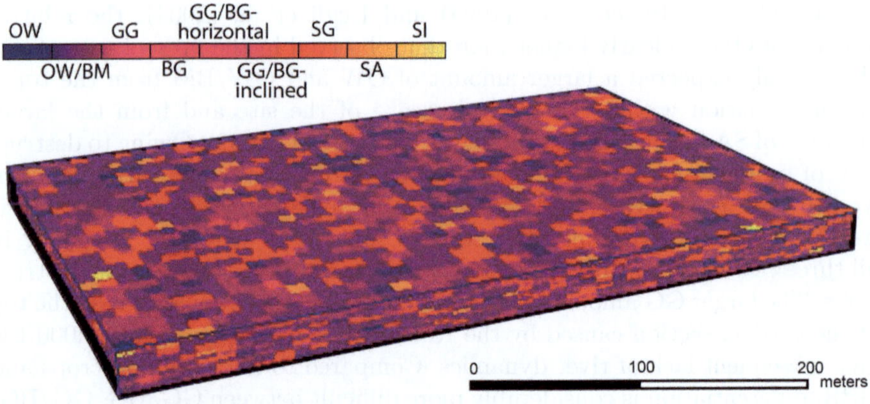

Fig. 7. Realization of sequential indicator simulation representing the distribution of sedimentary structure types identified at the investigation site. Refer to chapter 3.2 for the abbreviations

In each model run, the resulting probability density functions of the sedimentary structure types deviate less than ±10% from the initial probability density functions, which represent the expected volumetric fractions of the sedimentary structure types over the entire model domain. At least 100 or 1,000 runs are necessary to determine statistical moments and their confidence limits by Monte Carlo type modeling. In our case, the number of aquifer realizations was limited and the effects of subsurface heterogeneity along with changing boundary conditions illustrated qualitatively. A total of three different simulations were prepared, comprising different distributions of hydraulic conductivity, porosity and longitudinal dispersivity. These three simulations were again calculated for mean and distributed aquifer parameters, thus resulting in six realizations. These six realizations were then integrated into the small-scale groundwater model using three sets of hydrological and operational boundary conditions. For the three sets of boundary conditions, an equivalent uniform distribution of aquifer parameters over the entire model grid was also considered. Finally, a total of 18 simulations could be evaluated and compared.

4.3 Groundwater Modeling

Groundwater Flow Regime

Figure 8A-F illustrates the head distribution and flow paths for the three modeled boundary conditions, assuming uniform distribution of aquifer parameters (to the left) and for one of the simulations of the heterogeneously distributed aquifer parameters (on the right). The situation on March 23, 2003 (Fig. 8A&B) reflects the groundwater flow regime before the major construction phases and describes an ordinary situation. Note the backwater effect

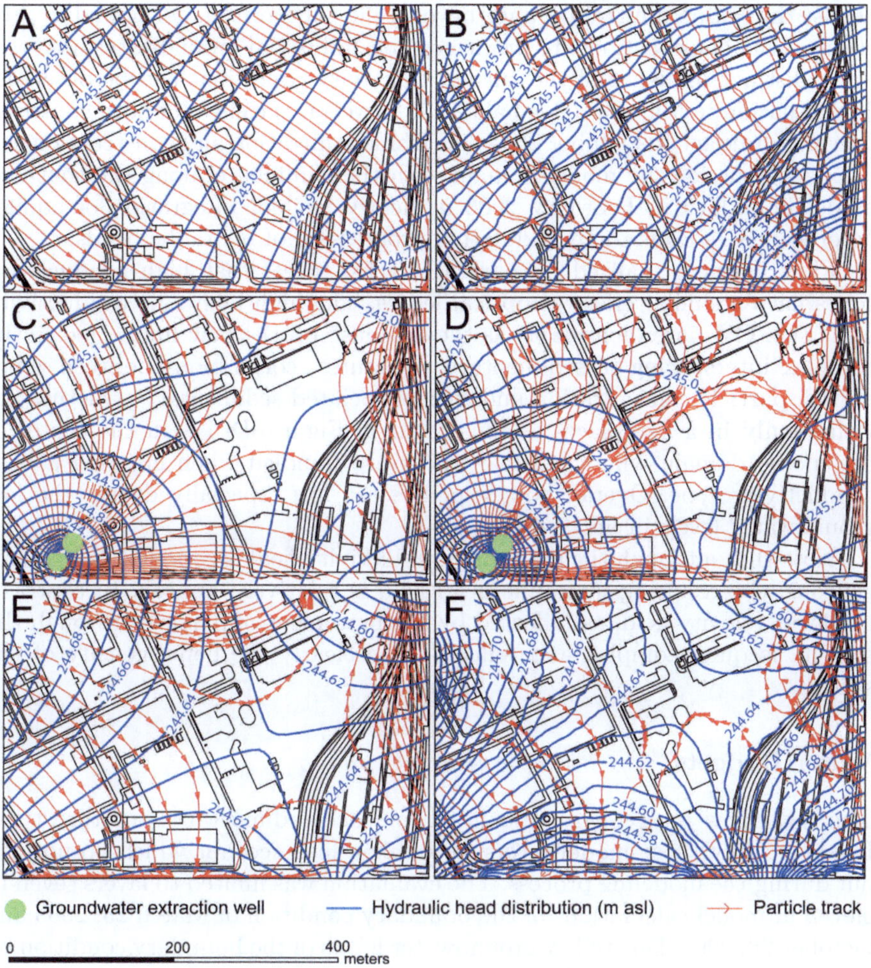

Fig. 8. Visualization of hydraulic head distributions (A, B, C: 0.05 m resolution; D: 0.1 m resolution; E, F: 0.01 m resolution) and flow paths illustrated by particle tracks (distance between two arrow heads indicates 100-day travel time) for three modeled situations for Layer 11. While the results for A, C and E are derived from a uniform distribution of aquifer properties, those for B, D and F are derived from a heterogeneous distribution of aquifer properties

behind the sheet pile wall and the diversion of flow paths. The situation on October 28, 2004 (Fig. 8C&D) reflects the groundwater flow regime during construction of the emergency exit. Note the inflow running towards the construction site drainage. The situation on February 16, 2006 (Fig. 8E&F) reflects the groundwater flow regime after completion of the emergency exit with the overall low groundwater levels. Note the low gradient of the hydraulic heads, the Rhine infiltration and, beyond the southern model boundary, the effect of the construction site drainages and industrial groundwater use.

A distinct influence of the current boundary conditions can be observed. The situation in March 2003 is balanced. However, the effect of both construction site drainage in October 2004 and Rhine infiltration and construction site drainages outside the model domain with their influence along the southern model boundary in February 2006 is clearly visible. Integration of aquifer heterogeneities leads to an undulating progression of hydraulic heads.

Flow paths, visualized by particle tracks, reveal that high conductivity zones have a similar effect as optical lenses. The high hydraulic conductivity units (i.e. OW) are identifiable as they "focus" on the flow lines. Due to the complex interspacing of sedimentary structures, correlations between zones of high particle concentration and their associated sedimentary structure are visible only in a few places as illustrated in Figure 8D, i.e. in the center of the model domain where particle tracks are bundled when entering a high conductivity zone. Overall flow velocities are low especially during the low groundwater levels in February 2006.

Note that although the hydraulic conductivity field is complex and heterogeneous, the resulting hydraulic head field is relatively smooth. In contrast, the local velocity field, as reflected by the movement of particles through the system, is quite complex and reflects more clearly the heterogeneity of the system.

Water Budgets

The total flow budget was calculated for each sedimentary structure type. However, only those model layers were taken into account which do not dry out during the modeling process. The evaluation was limited to layers seven to eleven in model calculations of the boundary condition of March 23, 2003 and October 28, 2004. Due to low groundwater levels of the boundary condition on February 16, 2006, the evaluation was restricted to layers eight to eleven only.

Figure 9 illustrates the relative amounts of flow budgets through the individual sedimentary structure types for one simulation of heterogeneously and uniformly distributed aquifer parameters. Note that the groundwater mainly flows through the GG and BG sedimentary structure types. As regards the heterogeneous distribution of aquifer parameters, the flow budgets of the more conductive sedimentary structure types OW and OW/BM are higher than those of the corresponding flow budgets of same cells, revealing uniform distribution of aquifer parameters. Also note that the three modeled boundary conditions show little influence on the overall distribution of the flow budgets.

Drill-core layer descriptions generally include mixed information of varying sedimentary structure types (Regli et al., 2002). Moreover, only a small number of indications of OW strata may often be identified in drill-core layer descriptions. Nevertheless, many outcrop observations reveal that OW and OW/BM occur frequently. Consequently, hydrogeological models based on

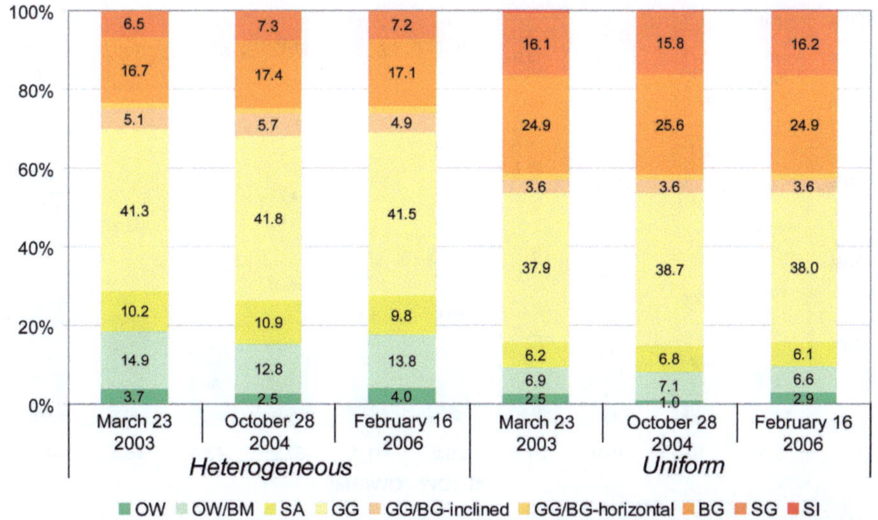

Fig. 9. Relative amounts of flow budgets through the individual sedimentary structure types diagramed for one aquifer heterogeneity simulation and for the uniform distribution of aquifer parameters. To facilitate comparison, relative flow budgets for the uniform distribution were calculated for the same cells as characterized by the sedimentary structure types for the heterogeneity simulation. Water budget values for the following sedimentary structure types were excluded in the graph: SI ranging between 0.1 and 0.5% and GG/BG-horizontal ranging between 1.1 and 1.6%. Refer to chapter 3.2 for the abbreviations

drill-core data may reproduce effective hydraulic conductivities, but they underestimate their standard deviations (Huggenberger and Regli, 2006). To assess the influence of larger amounts of sedimentary structure types OW and OW/BM, their relative amount was increased steadily. While the original simulations considered OW and OW/BM amounts of about 5.4%, the linear increase of these sedimentary structure types eventually totaled 54.0%. To allow a comparison of the different simulations and resulting flow budgets, the northern, fixed head model boundary was converted to fixed flow. Figure 10 illustrates the results of these simulations. By increasing OW and OW/BM amounts, a linear influence on the overall distributions of relative flow budgets was observed:

$$y = 6.05x + 13.61 \ (R^2 = 0.99). \tag{4.3.6}$$

However, considering merely OW, results in an increase in relative flow budgets capable of being described by a power law:

$$y = 2.74e^{0.24x} \ (R^2 = 0.97). \tag{4.3.7}$$

It should be noted, that these equations are adequate for this case study only and will differ from those of other case studies.

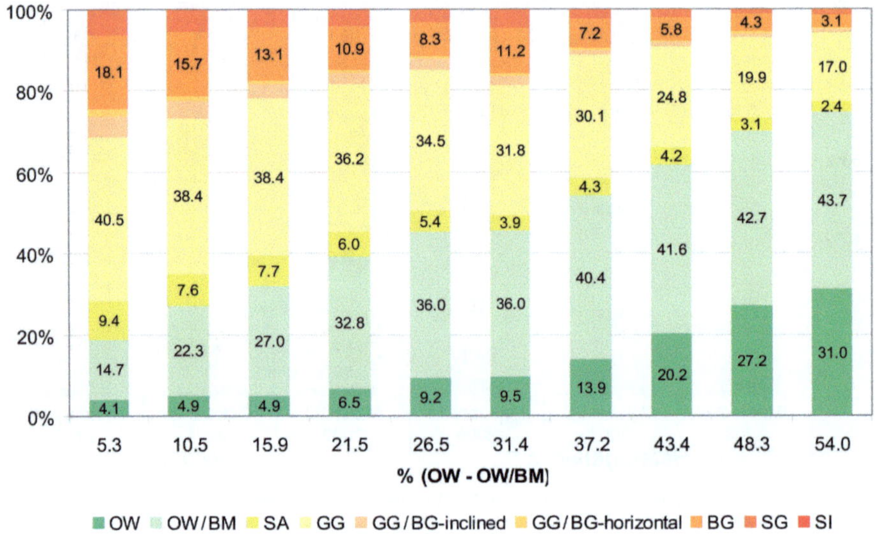

Fig. 10. Relative amounts of flow budgets through the individual sedimentary structure types diagrammed for the modeled boundary condition on March 23, 2003. The graph illustrates the distribution of flow budgets assuming a successive increase in relative abundance of the sedimentary structure types OW and OW/BM. Note that by increasing the relative amounts of water budgets for both OW and OW/BM, a linear influence is observed. However, an increase in the relative amounts of water budgets for OW alone results in an exponential influence. Water budget values for the following sedimentary structure types were excluded in the graph: SI ranging between 0 and 0.1%; SG ranging between 1.3 and 6.3%; GG/BG-horizontal ranging between 0.2 and 1.8% and GG/BG-inclined ranging between 1.0 and 5.0%. Refer to chapter 3.2 for the abbreviations

Flow Velocities

As a result of the high spatial variability of aquifer parameters, heterogeneities in the aquifer structure strongly affect transport behavior and also lead to a corresponding variability in the distribution of flow velocities through the individual sedimentary structure types. In fact, heterogeneity leads to small head variations, whereas those of velocities and travel time to large ones, as instinctively prognosticated when considering a layered aquifer with flow parallel to the bedding: although vertical heterogeneity does not produce any variations in the head distribution over the vertical, velocities may vary significantly from one layer to the next (de Marsily et al., 2005). The wide range of flow velocities is caused by uncertainty in hydraulic conductivity and flow paths. De Marsily et al. (1998) produced similar results but used particle arrival times to explain this effect.

The calculated average velocities v_x, v_y and v_z (see section 3.4.2) were evaluated for each sedimentary structure type. However, only those model layers were taken into account that do not dry out during the modeling process or do not comprise most of the inactive cells. The evaluation was thus restricted to layers eight to eleven. Figure 11 illustrates the distributions of flow velocities for boundary conditions on October 28, 2004.

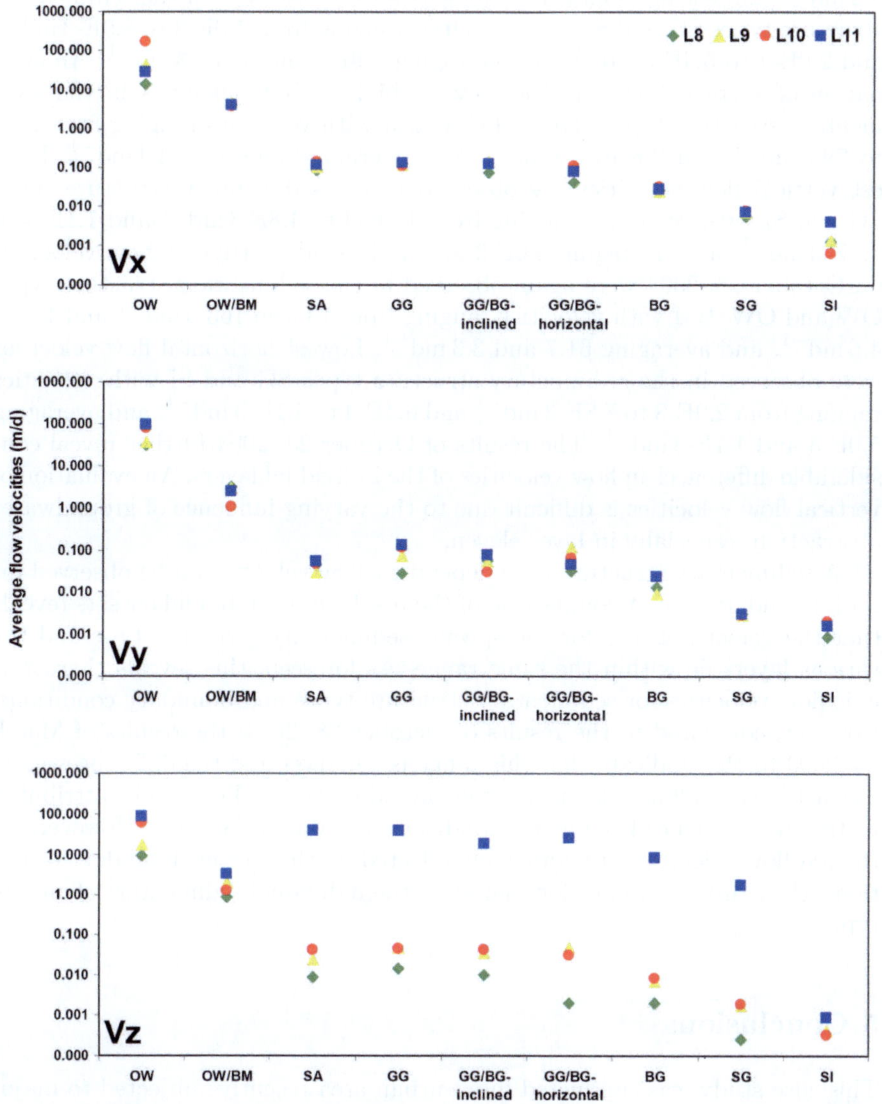

Fig. 11. Distributions of flow velocities for the boundary conditions on October 28, 2004. Note that the velocity values for the specific sedimentary structure types and the various layers lie within the same range, thereby revealing characteristic flow velocities for sedimentary structure types. Note also that the relative high vertical flow velocities in Layer 11 can be attributed to groundwater extractions. Refer to chapter 3.2 for the abbreviations

A comparison was established between the horizontal (v_x, v_y) and vertical (v_z) flow velocities for boundary conditions on March 23, 2003 and October 28, 2004. Highest average flow velocities on March 23, 2003 were observed in the sedimentary structure types OW and OW/BM with horizontal velocities ranging from 32.0 to 57.3 md^{-1} and 4.3 to 6.1 md^{-1}, and averaging 42.9 and 5.2 md^{-1}. Lowest horizontal flow velocities were observed in the sedimentary structure types SG and SI with velocities ranging from 7.0E-3 to 8.9E-3 md^{-1} and 2.0E-4 to 5.4E-3 md^{-1}, and averaging 7.9E-3 and 3.3E-3 md^{-1}. An evaluation of vertical flow velocities reveals highest flow velocities in the sedimentary structure types OW and OW/BM with velocities ranging from 22.8 to 79.6 md^{-1} and 2.6 to 5.8 md^{-1}, and averaging 50.8 and 4.4 md^{-1}. Lowest vertical flow velocities are observed in the sedimentary structure types SG and SI with velocities ranging from 1.8E-3 to 4.8E-3 md^{-1} and 1.1E-4 to 3.1E-3 md^{-1}, and averaging 3.6E-3 and 1.7E-3 md^{-1}. Highest flow velocities on October 28, 2004 were again observed in the sedimentary structure types OW and OW/BM with velocities ranging from 13.5 to 163.5 md^{-1} and 1.1 to 4.6 md^{-1}, and averaging 61.7 and 3.3 md^{-1}. Lowest horizontal flow velocities were observed in the sedimentary structure types SG and SI with velocities ranging from 2.9E-3 to 7.8E-3 md^{-1} and 6.4E-4 to 4.1E-3 md^{-1}, and averaging 5.0E-3 and 1.7E-3 md^{-1}. The results of October 28, 2004 further reveal considerable differences in flow velocities of the individual layers. An evaluation of vertical flow velocities is difficult due to the varying influence of groundwater abstraction, especially in layer eleven.

A sedimentary structure type-dependent flow velocity can be observed for both boundary sets. A comparison of the results of both boundary sets reveals that the velocity values for the specific sedimentary structure type and the various layers lie within the same range. As foreseen, this reveals characteristic flow velocities for sedimentary structure types and boundary conditions. However, compared to the results of October 28, 2004, the results of March 23, 2003 further indicate that this range is narrower and the differences more distinct between flow velocities of the individual layers. This can be attributed to the more balanced boundary conditions on March 23, 2003. However, although flow velocities are noticeably affected by the various boundary conditions, their distribution is dominated by the individual sedimentary structure type.

5 Conclusions

This case study was conducted in an urban area recently subjected to major changes in groundwater flow regime caused by subsurface tunnel road construction. Groundwater investigations generally focus on the required drawdown and dimensioning of construction site drainages. However, this approach is unsatisfactory as contamination is an additional factor to be considered in urban areas. To adequately evaluate potential mobilizations of contaminants, focus should be placed on aquifer heterogeneity. The presented methods, particularly their combination, exemplify quantitative data fusion as a practical tool for urban hydrogeology. The applied techniques allow integrating data of different quality into groundwater models. Furthermore, the potentials and limitations of this approach have also been identified. Since sedimentological information on the subsurface is often restricted to drill-cores, data on outcrops and geophysical surveys in urban areas is generally scarce. This may lead to one of the main constraints in geostatistical approaches, as it fails to derive connectivity properties of the sedimentary structures and generates disconnected ellipsoids of either high or low conductivity. Future work should focus on the distribution and connectivity of high-permeability sedimentary structure types (e.g. Proce et al., 2004).

The dynamics of the groundwater flow regime under changing spatial and temporal constraints were simulated and evaluated on the basis of a regional groundwater model. To prevent permanent negative impacts on the groundwater flow regime, particularly on groundwater quantity and quality, as well as irreversible deterioration of aquifer systems, recommendations for optimizing groundwater management are proposed and structural alternatives provided. Optimal dimension, operation and selection of sites appropriate for recharge wells and culverts are evaluated and their applicability verified. The modeling results are used to improve the groundwater monitoring system subsequently adapted to project requirements. Aside from managing various groundwater extraction and recharge operations, adequate groundwater protection (groundwater flow regime and quality) was also ensured. The groundwater management system also helped to identify changes in groundwater chemistry. Negative effects for industrial groundwater use could be minimized. So far, installation of supplementary recharge or interception wells was not necessary to ensure groundwater supply, or to prevent attracting contaminated groundwater. Simulation results reveal that the subsurface construction after completion does not significantly alter groundwater budgets or groundwater flow velocities.

Furthermore, as regards contaminant transport on a local scale, the applied techniques present an approach to quantify the effect of groundwater flow budgets and velocities in the individual hydrofacies. Obviously, groundwater flow in heterogeneous media occurs largely through interconnected highly permeable geological aquifer structures. Together with hydrological and operational boundary conditions they govern the groundwater flow and transport regime. However, the relative amounts of groundwater budgets through the individual hydrofacies do not appear to alter significantly for the various boundary conditions investigated. Moreover, single hydrofacies and their relative occurrence determine the distribution of groundwater budgets. Furthermore, an underestimation of the occurrence of highly permeable OW and OW/BM sedimentary structures and shift of relative amounts of groundwater budgets through the various sedimentary structure types was investigated.

When investigating contaminant transport, focus should be placed on relevant boundary conditions and origin, particularly on the depth of relevant substances. Modeling results reveal that the different sedimentary structures and simulated layers vary considerably in water budgets, flow paths and velocities. Investigations by Tompson and Gelhar (1990), Frind et al. (1988) and others reveal that heterogeneity dominates the movement of a contaminant plume at an early stage, and that the initial configuration of the plume influences its long-term evolution.

The outcrop in the investigation area clearly demonstrated the size of the relevant sedimentary structure types which are in the order of several tens to 100 m. A lithological description (nature and shapes of sedimentary structure) is required to describe as accurately as possible heterogeneity in sedimentary structure models and their resulting properties. Optimized acquisition of geological recording of drill-core data and less destructive drilling methods (drill cores in plastic liners) could significantly improve the characterization of sedimentary structure types. This includes more comprehensive hydrogeological investigations, i.e. a systematic collection and interpretation of drill-cores as a function of lithofacies as well as hydraulic and hydrogeochemical parameters. Such innovations could involve tailored exports from geological databases such as separation of sedimentary lithocomponents into light and dark-colored components. Since color variations are assumed to be an indicator of organic carbon content, they influence sorption capacities and sorption kinetics of the material. These informations are of prior importance considering groundwater transport processes.

However, numerous innovative technologies proposed for groundwater investigations face enormous implementation problems. As successful application is often questioned, conventional and more expensive approaches, such as extensive analytical programs, are favored. Applied and problem-oriented investigations should focus on a more sustainable management of groundwater resources in a sensitive urban environment. The described integrated approach, incorporating sedimentological and geostatistical analyses as well as groundwater modeling, may assist in meeting the challenges presented by such a sensitive urban environment and lead to more target-oriented remediation strategies. These include an evaluation of contaminated sites, risk assessment of waste disposal and parameterization of numerical groundwater models, thereby leading to the development of new approaches for complex practical problems. Although the results of this study are case-specific, the overall conceptual approach and methodologies used may be directly transferred to other urban areas.

Acknowledgments

We thank Sylvie Peter for reviewing the manuscript and two anonymous reviewers who provided valuable comments.

References

Adams EE, Gelhar LW (1992) Field study of dispersion in a heterogeneous aquifer, 2, Spatial moment analysis. Water Resour Res 28:3293–3308

Allen JRL (1978) Studies in fluviatile sedimentation: An exploratory quantitative model for the architecture of avulsion-controlled alluvial suits. Sed Geol 21:129–147

Anderson MP, Aiken JS, Webb EK, Mickelson DM (1999) Sedimentology and hydrogeology of two braided stream deposits. Sed Geol 129:187–199

Anderson MP, Woessner WW (1992) Applied groundwater modeling. Academic Press, San Diego

Anderson MP (1989) Hydrogeologic sedimentary structure models to delineate large-scale spatial trends in glacial and glaciofluvial sediments. Geological Society of America Bulletin 101:501–511

Ashmore P (1993) Anabranch confluence kinetics and sedimentation processes in gravel braided streams. In: Best JL, Bristow CS (eds) Braided Rivers, Geological Society Spec Publ 75:129–146

Ashmore P, Parker G (1983) Confluence scour in coarse braided streams. Water Resour Res 19:392–402

Ashworth PJ, Best JL, Jones M (2004) The relationship between sediment supply and avulsion frequency in braided rivers. Geol 32:21–24

Ashworth PJ, Best JL, Peakall J, Lorsong JA (1999) The influence of aggradation rate on braided alluvial architecture: field study and physical scale modelling of the Ashburton River gravels, Canterbury Plains, New Zealand. In: Smith ND, Rogers J (eds). Fluvial Sedimentology VI. Spec Publ Int Ass Sediment 28:333–346

Barrash W, Clemo T (2002) Hierarchical geostatistics and multifacies systems. Boise Hydrogeophysical Research Site, Boise, Idaho, Water Resour Res 38(10):1196

Boggs MJ, Young SC, Beard LM, Gelhar LW, Rehfeld, KR, Adams, EE (1992) Field study of dispersion in a heterogeneous aquifer. 1. Overview and site description. Water Resour Res 28(12):3281–3291

Bridge JS, Lunt IA (2006) Depositional models of braided rivers. In: Sambrook-Smith GH, Best JL, Bristow CS, Petts GE (eds). Braided Rivers: Process, Deposits, Ecology and Management. IAS Special Publication 36:11–50

Bridge JS (1993) The interaction between channel geometry, water flow, sediment transport and deposition in braided rivers. In: Best JL, Bristow CS (eds) Braided Rivers. Geological Society Spec Publ 75:13–71

Brierley GJ (1991) Bar sedimentology of the Squamish river, British Columbia: definition and application of morphostratigraphic units. J of Sed Petr 61:211–225

Carle SF, LaBolle EM, Weissmann GS, Van Brocklin D, Fogg GE (1998) Conditional simulation of hydrofacies architecture: A transition probability / Markov approach. In: Fraser GS, Davis JM (eds). Hydrogeologic Models of Sedimentary Aquifers. SEPM, Concepts in Hydrogeology and Environmental Geology, 1:147–170

Carrera J (1993) An overview of uncertainties in modeling groundwater solute transport. J of Con Hyd 13(1–4):23–48

Chiang W-H (2005) 3D-Groundwater Modeling with PMWIN – A simulation system for modeling groundwater flow and transport processes. 2nd edn Springer, Berlin, 397pp

Church M (2002) Geomorphic thresholds in riverine landscapes. First international symposium on landscape dynamics of riverine corridors, Ascona, Switzerland, March 2001. Freshwat Biol 47:541–557

Dagan G (2002) An overview of stochastic modelling of groundwater flow and transport. From theory to applications. EOS (Trans Am Geophys Soc), 83/53:621–625

de Marsily G, Delay F, Gonçalvès J, Renard P, Teles V, Violette S (2005) Dealing with spatial heterogeneity. Hydrogeol J 13:161–183

de Marsily G, Schafmeister MT, Delay F, Teles V (1998) On some current methods to represent the heterogeneity of natural media in hydrogeology. Hydrogeol J 6:115–130

Deutsch CV, Journel AG (1998) GSLIB: Geostatistical software library and user's guide, 2nd edn Oxford University Press, Oxford, 369pp

Deutsch CV, Wang L (1996) Hierarchical object-based stochastic modelling of fluvial reservoirs. Math Geol 28:857–880

Environmental Modeling Systems Inc. (2006) GMS, groundwater modeling system v6.0. http://www.ems-i.com

Flynn R (2003) Virus transport and attenuation in perialpine gravel aquifers. PhD. Thesis, University of Neuchâtel, Switzerland, 178pp

Fogg GE, Noyes CD, Carle SF (1998) Geologically based model of heterogeneous hydraulic conductivity in an alluvial setting. Hydrogeol J 6:131–143

Frind EO, Sudicky EA, Schellenberg SL (1988) Micro-scale modelling in the study of plume evolution in heterogeneous media, In: Custodio E et al. (eds), Groundwater Flow and Quality Modelling. Reidel, Dordrecht: 439–461

Gelhar LW (1986) Stochastic subsurface hydrology from theory to applications. Water Resour Res 22:135S–145S

Graf H, Schäfer W (2002) Simulation des Nitrattransports im Einzugsgebiet eines Wasserwerks. Grundwasser 4/2002:233–242

Heller P, Paola C (1992) The large-scale dynamics of grain-size variation in alluvial basins, 2: Application to syntectonic conglomerate. Basin Res 4:73–90

Hill MC, Banta ER, Harbaugh AW, Anderman ER (2000) MODFLOW-2000, the U.S. Geological Survey. Modular Ground-Water Model—User guide to the observation, sensitivity, and parameter-estimation processes and three post-processing programs: U.S. Geological Survey Open-File Report 00-184, 210 pp

Huggenberger P, Regli C (2006) A sedimentological model to characterise braided river deposits for hydrogeological applications. In: Sambrook-Smith GH, Best JL, Bristow CS, Petts GE (eds). Braided Rivers: Process, Deposits, Ecology and Management. IAS Special Publication 36:51–74

Huggenberger P, Epting J, Spottke I, Regli C, Zechner E (2006) Fluss-Grundwasser-Interaktion, Interreg III Projekt A MoNit: Modellierung der Grundwasserbelastung durch Nitrat im Oberrheingraben, Landesamt für Umwelt, Messungen und Naturschutz Baden-Württemberg, Karlsruhe

Huggenberger P, Aigner T (1999) Introduction to the special issue on aquifer-sedimentology: problems, perspectives and modern approaches. Sedimentary Geology 129:179–186

Huggenberger P, Siegentaler C, Stauffer F (1988) Grundwasserströmung in Schottern: Einfluss von Ablagerungsformen auf die Verteilung der Grundwasserfliessgeschwindigkeit. Wasserwirtschaft 78:202–212

Journel AG, Huijbregts CJ (1989) Mining Geostatistics. Academic Press, New York

Jussel P, Stauffer F, Dracos T (1994) Transport modeling in heterogeneous aquifers: 1. statistical description and numerical generation of gravel deposits. Water Resour Res 30(6):1803–1817

Kirchhofer RT (2006) GeoData. Geological Data-base of the city of Basel and northwestern Switzerland

Klingbeil R, Kleineidam S, Asprion U, Aigner T, Teutsch G (1999) Relating lithofacies to hydrofacies structure: outcrop-based hydrogeological characterisation of Quaternary gravel deposits. Sed Geol 129:299–310

Koltermann CE, Gorelick SM (1996) Heterogeneity in sedimentary deposits: A review of structure-imitating, process-imitating, and descriptive approaches. Water Resour Res 32:2617–2658

Mackay DM, Freyberg DL, Roberts PV, Cherry JA (1986) A natural gradient experiment on solute transport in a sand aquifer. 1. Approach and overview of plume movement. Water Resour Res 22(13):2017–2029

Maxwell RM, Pelmulder SD, Tompson AFB, Kastenberg WE (1998) On the development of a new methodology for groundwater-driven health risk assessment. Water Resour Res 34(4): 833–847

Maxwell RM, Kastenberg WE (1999) Stochastic environmental risk analysis: an integrated methodology for predicting cancer risk from contaminated groundwater. Stochastic Environmental Research and Risk Assessment 13:27–47

McDonald MG, Harbaugh AW, Banta ER, Hill MC (2000) Modflow-2000. The U.S. geological survey modular ground-water model-user guide to modularization concepts and the ground-water flow process. US Geological Survey Open File Report 00-92

McKenna SA, Poeter EP (1995) Field example of data fusion in site characterization. Water Resour Res 31:3229–3240

Miall AD (1985) Architectural-element analysis: A new method of sedimentary structure analysis applied to fluvial deposits. Earth-Science Rev 22:261–308

Miall AD (1978) Lithofacies types and vertical profile models in braided rivers: a summary. In: Miall AD (ed), Fluvial Sedimentology. Can Soc Pet Geol Mem 5:605–625

National Research Council (1990) Groundwater models: Scientific and regulatory applications. National Academy Press, Washington, DC

Nowak W, Cirpka OA (2006) Geostatistical inference of hydraulic conductivity and dispersivities from hydraulic heads and tracer data, Water Resour Res 42(8)

Paola C, Heller P, Angevine C (1992) The large-scale dynamics of grain-size variation in alluvial basins, 1: Theory. Basin Res 4:73–90

Proce CJ, Ritzi RW, Dominic DF, Dai Z (2004) Modeling multiscale heterogeneity and aquifer interconnectivity. Ground Water 42(5):658–670

Rauber M, Stauffer F, Huggenberger P, Dracos T (1998) A numerical three-dimensional conditioned/unconditioned stochastic sedimentary structure type model applied to a remediation well system. Water Resour Res 34(9):2225–2233

Regli C, Rosenthaler L, Huggenberger P (2004) GEOSSAV: a simulation tool for subsurface applications. Computers & Geosciences 30:221–238

Regli C, Rauber M, Huggenberger P (2003) Analyses of aquifer heterogeneity within a well capture zone, comparison of model data with field experiments: A case study from the river Wiese, Switzerland. Aquatic Sciences 65:111–128

Regli C, Huggenberger P, Rauber M (2002) Interpretation of drill-core and georadar data of coarse gravel deposits. J of Hydrol 255:234–252

Rehfeldt KR, Boggs JM, Gelhar LW (1993) Field study of dispersion in a heterogeneous aquifer, 3, Geostatistical analysis of hydraulic conductivity. Water Resour Res 28:3309–3324

Rentzel P (1994) Geologisch-bodenkundliche Untersuchungen an den Niederterrassenfeldern bei Basel unter besonderer Berücksichtigung der späteiszeitlichen Fundstelle Basel-Gasfabrik, Jahresbericht der Archäologischen Bodenforschung des Kantons Basel-Stadt:31–52

Rubin Y, Lunt IA, Bridge JS (2006) Spatial variability in river sediments and its link with river channel geometry. Water Resour Res 42(6)

Schulze-Makuch D, Carlson DA, Cherkauer DS, Malik P (1999) Scale dependency of hydraulic conductivity in heterogeneous media. Ground Water 37(6):904–919

Siegenthaler C, Huggenberger P (1993) Pleistocene Rhine gravel: deposits of a braided river system with dominant pool preservation. In: Best JL, Bristow CS (eds) Braided Rivers. Geological Society Spec Publ 75:147–162

Sudicky EA (1986) A natural gradient experiment on solute transport in a sand aquifer: Spatial variability of hydraulic conductivity and its role in the dispersion process. Water Resour Res 22(13):2069–2082

Teles V, Delay F, de Marsily G (2004) Comparison of genesis and geostatistical methods for characterizing the heterogeneity of alluvial media: groundwater flow and transport simulations. J Hydrol; 294(1–3):103–121

Teutsch G, Kobus H (1990) The environmental research field site "Horkheimer Insel:" Research Program, instrumentation and first results. J Hydraulic Res 28(4):491–501

Tompson AFB, Gelhar LW (1990) Numerical simulation of solute transport in three-dimensional randomly heterogeneous porous media. Water Resour Res 26(10):2541–2562

Wagner U, Huggenberger P, Schaub D, Thater M (2001) Interreg III, Erkundung der Grundwasserleiter und Böden im Hochrheintal. Herausgeber: Landratsamt Waldshut, pp101

Webb EK, Anderson MP (1996) Simulation of preferential flow in three-dimensional, heterogeneous conductivity fields with realistic internal architecture. Water Resour Res 32:533–545

Webb EK (1994) Simulating the three-dimensional distribution of sediment units in braided-stream deposits. J Sediment Res B64:219–231

Weissmann GS, Carle SF, Fogg GE (1999) Three-dimensional hydrofacies modeling based on soil surveys and transition probability geostatistics. Water Resour Res 35(6):1761–1770

MIX
Papier aus verantwortungsvollen Quellen
Paper from responsible sources
FSC® C105338

If you have any concerns about our products,
you can contact us on
ProductSafety@springernature.com

In case Publisher is established outside the EU,
the EU authorized representative is:
Springer Nature Customer Service Center GmbH
Europaplatz 3, 69115 Heidelberg, Germany

Printed by Libri Plureos GmbH
in Hamburg, Germany